VERTEBRATE
HARD TISSUES

THE WYKEHAM SCIENCE SERIES

General Editors:

PROFESSOR SIR NEVILL MOTT, F.R.S.
Emeritus Cavendish Professor of Physics
University of Cambridge

G. R. NOAKES
Formerly Senior Physics Master
Uppingham School

Biological Editor:

W. B. YAPP
Formerly Senior Lecturer
University of Birmingham

The aim of the Wykeham Science Series is to introduce the present state of the many fields of study within science to students approaching or starting their careers in University, Polytechnic, or College of Technology. Each book seeks to reinforce the link between school and higher education, and the main author, a distinguished worker or teacher in the field, is assisted by an experienced sixth form schoolmaster.

VERTEBRATE

HARD TISSUES

L. B. Halstead
University of Ife. Nigeria

WYKEHAM PUBLICATIONS (LONDON) LTD
(A MEMBER OF THE TAYLOR & FRANCIS GROUP)
LONDON AND WINCHESTER
1974

First published 1974 by Wykeham Publications (London) Ltd.

© *1974 L. B. Halstead. All rights reserved.*

Cover illustration—Bone, tooth and horn of ram by Jennifer Middleton

ISBN 0 85109 430 9

Printed in Great Britain by Taylor & Francis Ltd.
10–14 Macklin Street, London, WC2B 5NF

Distribution:

UNITED KINGDOM, EUROPE, MIDDLE EAST AND AFRICA

Chapman & Hall Ltd. (a member of Associated Book Publishers Ltd.), 11 North Way, Andover, Hampshire.

WESTERN HEMISPHERE

Springer-Verlag New York Inc., 175 Fifth Avenue, New York, New York 10010.

AUSTRALIA, NEW ZEALAND AND FAR EAST (excluding Japan)

Australia and New Zealand Book Co. Pty. Ltd., P.O. Box 459, Brookvale, N.S.W. 2100.

ALL OTHER TERRITORIES

Taylor & Francis Ltd., 10–14 Macklin Street, London, WC2B 5NF.

PREFACE

VERTEBRATE hard tissues include a multitude of materials from bones and teeth to horn and eggshells. The tools used to study them are equally numerous, from those of the biochemist to the engineer, taking in the approaches of histologists, anatomists, pathologists and palaeontologists on the way. In this monograph I begin with the molecular and cellular aspects and then discuss the different tissues separately. Problems relating to the skeletal system as a whole are dealt with and the book ends with an abbreviated account of trauma and disease. My aim has been to trace the study of vertebrate hard tissues through their different levels of organization and to demonstrate the essential unity of knowledge so obtained, for no single discipline can provide solutions. I have been at particular pains to emphasize throughout the importance of comparative studies among the vertebrates, with especial stress being placed on the concept of changes through time—in other words evolution. Evolution occurs at all levels, at the molecular, cellular, tissue, organ, whole animal and community level. Strangely enough, it is only in the last few decades that attention has been paid to the evolutionary process at the molecular and lower levels of animal organization.

Even though vertebrate hard tissues are the subject of study from the standpoint of many disciplines, these tend to be isolated from one another by barriers of different technical languages—less politely termed jargon. There is an abundance of advanced texts in which particular topics are considered in great depth, but as far as I am aware nobody has yet attempted an overview covering the entire spectrum at an elementary level.

I am deeply conscious of Dame Janet Vaughan's admonition that 'it is certainly presumptuous for one individual to attempt to survey a field which today occupies the attention of specialists from many different disciplines'. Nonetheless, I trust that this first attempt will at least serve as an introduction to the study of vertebrate hard tissues. But more important, I hope this book will encourage students to delve deeper or even eventually to enter the expanding field of calcified tissue research.

L. B. HALSTEAD

Royal Dental Hospital, London
September 1973.

For Jenny

ACKNOWLEDGMENTS

DURING the preparation of this book I have been fortunate to receive constructive and oft-times trenchant criticism from my colleagues Professor W. G. Armstrong, Dr G. Embery, Miss Angela Gilhespie, Dr J. S. Greenspan, Dr Margaret Hackemann, Dr A. F. Hayward, Dr Karen Hiiemae, Dr T. S. Kemp, Mr N. Priest and Dr F. B. Reed. I have gained much over the years from discussions with Mrs Cherrie Bramwell, Dr T. Ørvig, Professor F. G. E. Pautard and Professor K. Simkiss. Photographs were generously provided by Dr A. Boyde (scanning electron micrographs), Dr B. Boothroyd (electron micrograph of osteoclast brushborder), Electron Microscopy Unit of the Royal Dental Hospital headed by Dr. A. F. Hayward (electron micrographs) and Mr J. R. Mercer (photomicrographs).

Miss Jennifer Middleton has provided, with her usual expertise, line illustrations which greatly enhance any merit this work may have; without her constant unflagging encouragement over several years and continents the text itself would never have been completed.

Most of the writing was undertaken at the Royal Dental Hospital and I am grateful to Professors R. B. Lucas, W. G. Armstrong and H. J. J. Blackwood for their hospitality. The bulk of the typing was by Mrs Sandra Collo and Miss Remi Oluwatimilehin. Finally I must express my thanks to Mr R. Hill for his patience and ever-constructive criticisms as the manuscript evolved.

As several, if not all, of the above ladies and gentlemen disagree strongly in some part or other with my views, I hereby both severally and collectively absolve them from any responsibility for views both heretic and erroneous, which I have expounded.

CONTENTS

x

PART I

MOLECULAR AND CELLULAR ASPECTS

CHAPTER 1
protein evolution

1.1. *Protein synthesis*

THE evolution of the vertebrates has been carefully documented from fossil bones and teeth. Gradual changes in the skeleton and dentition can be related to changes in function which can be further related to the environments in which the animals lived. At the same time it has long been known that the genetic material of an organism is contained in the chromosomes in the nucleus. Moreover, there was no evidence that the genes of an animal could be affected directly by changes in the environment. Rather 'random changes' or mutations of the genes occurred and natural selection of these gradually spread them through the population or eliminated them, depending on whether they conferred an advantage or the reverse.

The nature of the gene was something that eluded biologists, but during the 1930's and 1940's evidence was accumulated which suggested that the answer lay in the nucleic acids and deoxyribonucleic acid (DNA) in particular. Deoxyribonucleic acid is a polymer made up of nucleotides. These comprise one of four bases, attached to the sugar molecule deoxyribose which is attached to a phosphate group. The nucleotides are linked via the phosphate groups. The alternating phosphate and sugar groups form the backbone to which are attached purine and pyrimidine bases. It had been noted that the proportions of the bases in DNA are such that the purine and pyrimidines always seem to match. Hence adenine (a purine) paired with thymine (a pyrimidine) and guanine (a purine) with cytosine (a pyrimidine) (fig. 1.1*b*). This evidence led Watson and Crick in 1953 to propose a double helix structure for DNA (fig. 1.1*a*). They suggested that DNA consisted of two complementary strands that were twisted round one another. When separated, each individual strand would form the template for a new complementary strand. In one of the most famous understatements in science they concluded 'it has not escaped our notice that the specific pairing we have postulated immediately suggests a possible copying mechanism for the genetic material'.

The DNA molecule, located in the nucleus, controls the mechanism of heredity. Clearly the genetic information has to be carried from the nucleus to the site of protein synthesis, which is in the cytoplasm. An enzyme RNA-polymerase opens up the double helix of DNA. On the exposed bases a further molecule is formed of complementary bases, except that instead of thymine the pyrimidine uracil occurs, and the sugar is ribose (fig. 1.1c). This molecule, termed messenger ribonucleic acid (mRNA), separates from the DNA molecule and moves to the cytoplasm. The DNA is said to transcribe for mRNA. This is the messenger which carries the genetic information from the nucleus to the cytoplasm.

Messenger RNA forms the assembly line on which amino acids become linked together to form peptides. The sequence of bases on the mRNA molecule is the genetic code. If each base coded for a single amino acid then there could be proteins made up of only four (4^1) types of amino acid. If a combination of two bases coded for an amino acid, a maximum of 16 ($=4^2$) could be formed. However, some twenty amino acids are known and this means that at least three bases must code for an amino acid. Such a triplet code will have 64 ($=4^3$) 'words'. To continue the analogy of language, there are also triplets which act as capital letters and full stops. As there are clearly many more triplets than the twenty amino acids this type of code is described as degenerate. In fact several triplets code equally for the same amino acid.

Once the message of the gene has been transcribed from DNA to mRNA there is then the problem of actually translating this into the appropriate amino acid sequence. A further molecule known as transfer RNA (tRNA) acts as an "adaptor". Each amino acid is bound to its particular tRNA by a specific enzyme, the complete unit being known as amino-acyl–tRNA. The tRNA molecule is thought to be somewhat clover-leaf shaped and one of the "leaves" bears three bases, the anti-codon, which complement the appropriate triplet codon on the mRNA molecule. The tRNA molecules carrying the amino acids plug into the appropriate slots on the mRNA molecule (fig. 1.1d).

The actual linking of the amino acids is by the formation of peptide bonds in which the amino group of one amino acid is joined to the carboxyl of the adjacent amino acid with the loss of a molecule of water (fig. 1.2). This process is accomplished in cytoplasmic organelles, the ribosomes, by enzymes. The ribosomes become attached to the mRNA and move along the molecule occupying two codon sites, the peptide site and the amino acid site. The amino acid site is where the tRNA carrying its amino acid plugs into the appropriate mRNA codon; the peptide site is where the previous tRNA was plugged in but where now a peptide bond is formed with the new recently arrived amino acid. The tRNA now relieved of its burden returns to the cytoplasm for a further load. As more amino acids are added on, a peptide chain is formed which is eventually released into the cytoplasm as a protein molecule (fig. 1.1).

The sequence of amino acids is described as the primary structure, and this largely determines the eventual configuration of the complete protein molecule. By studying such sequences it is possible to determine

2

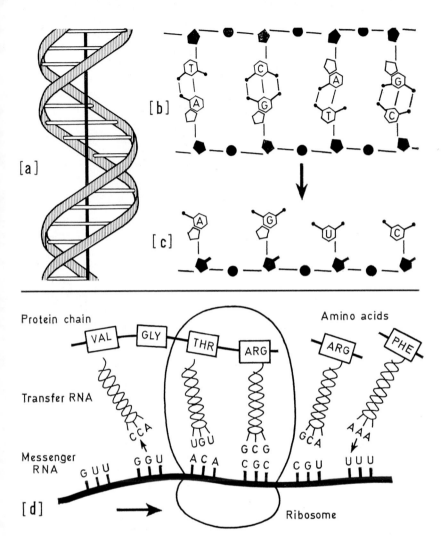

Fig. 1.1. (a) Double helix of deoxyribonucleic acid (DNA), with structure similar to a twisted ladder, the base pairs forming the rungs. (b) Diagram of DNA to show pairing of bases, thymine with adenine and guanine with cytosine. The backbone of each DNA strand consists of deoxyribose sugar alternating with a phosphate molecule. (c) Transcription of DNA to messenger ribonucleic acid (mRNA), in which the base uracil replaces thymine and the sugar is ribose. (d) Translation of sequence of bases on mRNA to sequence of amino acids. Ribosome within which protein synthesis takes place moves along mRNA, transfer or tRNA carrying a specific amino acid recognizes the appropriate triplet codon on the mRNA and plugs into the amino acid site and as the ribosome moves onward the amino acid joins the previous one. This position is termed the peptide site. When thus relieved of its burden the tRNA returns to the cytoplasm for a further amino acid.

3

the genetic code. By further comparing the primary structure of related molecules from different organisms it becomes possible to suggest the outline of the course of evolution at the molecular level.

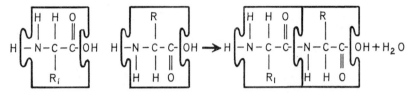

Fig. 1.2. Diagram of peptide bond. Amino acids are linked by the amino (N terminal) group of one attaching to the carboxyl (C terminal) group of the next with the elimination of a molecule of water, forming the peptide (–CO–NH–) bond (after Clowes).

1.2. *Evolution of haemoglobin*

Few studies have so far been undertaken on the evolution of proteins. Perhaps the best authenticated example so far is that of the respiratory pigment haemoglobin, which has been studied by E. Zuckerkandl and L. Pauling. In no sense is haemoglobin related to hard tissues, but it can usefully be considered as an illustration of a method of obtaining important evolutionary information. Each haemoglobin molecule consists of four polypeptide chains which together enfold a haem group, which is concerned with carrying oxygen and carbon dioxide to and from the tissues. There are different polypeptide chains, which are designated by the letters of the Greek alphabet. In man there are two pairs of polypeptide chain, the alpha and beta; in some cases there is a pair of delta instead of beta. Before birth the foetal haemoglobin comprises a pair of alpha chains and a pair of gamma chains. In spite of the differences of these chains the overall shape of the haemoglobin molecule—the tertiary structure—does not seem to vary. Nevertheless, there are considerable differences in their respective amino acid sequences. For example in man the alpha and beta chains have 64 identical sites and 77 different ones.

The beta chain differs from the gamma chain at 39 sites and from the delta chain at only 10. The alpha chain differs from the myoglobin chain at about 110 sites. Myoglobin is the respiratory pigment of muscle and has a tertiary structure similar to that of haemoglobin.

From these differences in the primary structures of the various polypeptide chains, it may be suggested that all these molecules have a common origin, and that the greater the number of differences the greater the antiquity of the common ancestor. Conversely, the fewer the differences the closer the genetic relationship.

These comparisons are between different chains within a single species —man. At the same time, it is possible to compare the same chain in different species, and so be able to say something about probable interspecies relationships. For example, the gorilla beta chain differs from the human at only one site, the pig differs at 17 sites. This suggests that man is more closely related to the gorilla than to the pig, a view which is in conformity with evidence of gross anatomy.

From this kind of evidence Zuckerkandl and Pauling calculated, from the generally accepted fact that the common ancestor of modern placental mammals was about 80 million years ago, that it took about 7 million years for an effective amino acid substitution to take place. For very large differences they suggested a period of 10 million years on the grounds that multiple mutations would be required. If the rationale of this approach is accepted, an evolutionary tree for the evolution of the different chains of the haemoglobin molecule can be drawn up and superimposed on the geological time-scale.

Taking the Zuckerkandl and Pauling scheme as a working hypothesis, it transpires that the timing of the origins of the different polypeptide chains, including the alpha chain from myoglobin, appear to coincide with major events in the history of the vertebrates. The delta chain, which is known only among the anthropoids, i.e. the higher primates, comes out at 35 million years, the beginning of the Oligocene period just after the first record of the anthropoids from the Eocene rocks of Egypt. The origin of the beta chain at 150 million years is roughly at the time when it is believed the placental mammals may have originated. The gamma chain at 380 million years is the point at which the crossopterygian fish are first known from the fossil record. This is the group that gave rise to all the land vertebrates. The divergence of the haemoglobin molecule from that of myoglobin comes at 650 million years in the Precambrian, where there is no record of the vertebrates, but at which time it is conventionally postulated they must have first arisen! (fig. 1.3).

The concern of the molecular biologists working in this field has been to trace back from the present a line of descent. V. M. Ingram suggested that the points of postulated common origin must have involved the duplication and translocation of the genes for the different polypeptide chains of haemoglobin. Once this had happened, mutations would take place independently in the different genes and hence differences would develop resulting in the divergence of the chains from one another. The concept of gene duplication and translocation is the only reasonable explanation for the existence of the different chains.

Such chromosomal alterations are obviously of major importance compared with mutations involving amino acid substitutions at particular loci along a polypeptide chain. That they appear to coincide with major events in the history of the vertebrates may be merely a coincidence. On the other hand, it may well be that major evolutionary stages at the molecular level are connected with those at the somatic level.

The four stages in the postulated evolutionary scheme of the haemo-

5

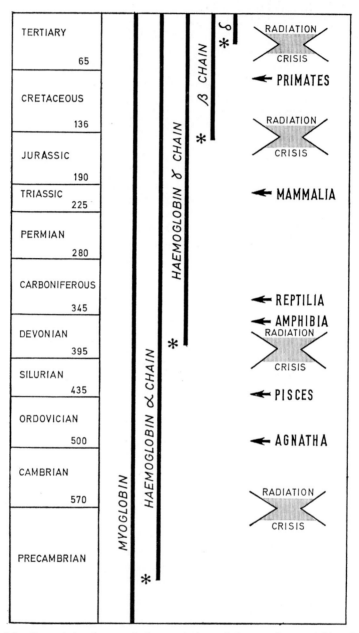

Fig. 1.3. Suggested scheme of the evolution of human haemoglobin chains through geological time. Origins of major vertebrate taxa indicated, together with selected periods of crisis and radiation. Asterisks represent points in time where gene duplication and translocation are believed to have occurred. Figures give approximate time in millions of years from beginning of geological periods.

globin molecule seem also to coincide with major crises. At the end of the Precambrian there was a world-wide ice-age and many of the organisms known from the seas at the time became extinct. Thereafter, there was an explosion of animal life and the ability to lay down a skeleton was acquired in many unrelated phyla. The next crisis was at the end of the Lower Devonian period when most of the dominant fresh-water vertebrates died out and there began an evolutionary radiation of both the cartilaginous and bony fishes. The third crisis is more difficult to recognize. During the age of Dinosaurs there were many groups of mammals, almost all of which died out by the end of the Jurassic. There-after, during the Cretaceous, the ancestors of the modern mammals, including the first primates, evolved. The final crisis was at the beginning of the Tertiary Period when the primates, which had filled the ecological niche of the rodents, were replaced by genuine rodents. After the virtual eclipse of the primates, there began a further development, which led eventually to man.

The origin of the different haemoglobin chains, which involved the duplication and translocation of genes, coincided with periods of crisis followed by radiations of the survivors.

It has recently been discovered by geneticists that gene duplication and translocation can occur during the intense inbreeding that takes place during bottle-neck situations, when there is a drastic fall in numbers of a population.

It may well be that the evolutionary radiations following crises, which are a recognized pattern of earth history, have played an important role in evolution at the molecular level.

1.3. *Evolution of collagen*

There is only one protein, the fibrous protein collagen, that seems to have survived the vicissitudes of geological time. If, some time in the future, molecular biologists are able to put forward an evolutionary scheme for collagen, it might be possible to check the results from the fossil record. Collagen is the major organic component of bones and teeth, and it has been known for over a decade that it survives in fossil bones and teeth and is amenable to biochemical analyses. The prob-lems involved in interpreting the results obtained from fossil material include contamination from the enclosing sediment and the differential breakdown of the material, as well as evolution. The problem is how to disentangle these different aspects. To date, this has not been accom-plished, but a number of advances have already been made.

J. Pikkarainen and E. Kulonen in Finland have studied the different amino acid composition of the three polypeptide chains that make up the collagen molecule in different animals, and have put forward a scheme which requires two successive duplications and translocations of the collagen genes. These are at the point where the early vertebrates

7

gave rise to the advanced poikilotherms or cold-blooded vertebrates, and the latter to the homoiothermic (warm-blooded) vertebrates, the birds and mammals (fig. 1.4). These two events would be placed in the geological record at the beginning of the Devonian and in the middle of the Mesozoic Era respectively. Roughly this coincides with two of the major events of the evolution of the haemoglobin molecule.

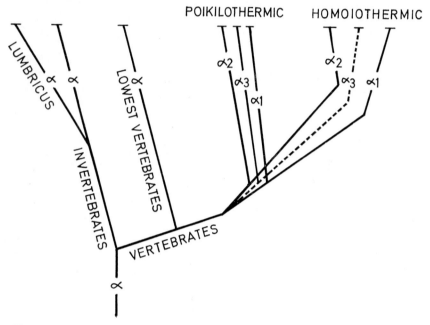

Fig. 1.4. Suggested evolution of collagen polypeptide chains (after Pikkarainen and Kulonen).

Unfortunately the complete primary structure of collagen is not fully known, so that it is impossible to draw up a detailed evolutionary scheme like that put forward for haemoglobin. The structure of collagen will be discussed in the next chapter, but for the present it is worth noting that the collagens of birds and mammals contain less of the amino acids threonine and serine, but more proline, than those of cold-blooded vertebrates such as frogs and bony and cartilaginous fishes. The total amount of these three amino acids is constant between the two groups, and threonine and serine are replaced by proline. The mRNA codons for threonine (ACU, ACC, ACA, ACG) and serine (UCU, UCC, UCA, UCG) only differ from that for proline (CCU, CCC, CCA, CCG) in the first letter of the codon. A single mutation affecting the first base of the triplet on the DNA molecule will result in the new triplets coding for a different amino acid, and in this case threonine and serine would be gradually substituted by proline. If this mutation is advantageous, it would tend to be favoured by natural selection.

Although there are not yet enough data available from analyses of fossil collagens, there is already a suggestion that the amounts of serine and threonine relative to proline are fairly large. At the present state of knowledge it can only be said that the evidence from the fossil record is not inconsistent with the hypothesis put forward by Pikkarainen and Kulonen.

It is clear that collagen has undergone evolutionary changes by the gradual substitution of certain amino acids by others at particular loci. Furthermore, as with haemoglobin, it is necessary to postulate major mutations involving the duplication and translocation of the genes responsible for the collagen polypeptide chains. From the distribution of the types of collagen among the vertebrates, gene duplication seems to have coincided with major events in vertebrate evolution.

CHAPTER 2

calcifying macromolecules

2.1. *Collagen*

Collagen is a fibrous protein which is a major component of connective tissues throughout most of the animal kingdom. In all vertebrates it forms the organic matrix with which calcium phosphate is associated to produce the tissues bone and dentine. As well as this, collagen is found throughout the body. Tendons comprise some 90 per cent collagen dry weight whereas the vitreous humour of the eye contains only 0·05 per cent. This protein performs a variety of functions. The sclera and the cornea of the eye are collagenous. The transparency of the cornea is due to the arrangement of the fibrils, which in turn is dictated by the mode of attachment of carbohydrate moieties to collagen. If an eyeball is squeezed so that this arrangement is disturbed the cornea becomes opaque. The mechanical properties of certain tissues are a consequence of their collagenous framework. Alterations of the mechanical properties of collagen can take place very rapidly. The cervix of the uterus reveals a remarkable plastic extensibility of its collagenous framework during birth of the young. Similarly the foetal membranes decrease their tensile strength as the embryo develops.

Collagen fibrils are made up of aggregations of the tropocollagen macromolecules. These are the basic structural units of collagen. Each consists of three polypeptide chains, the secondary structure of each chain being a left-handed helix. In cartilage these are believed to be identical and are alpha 1 chains. In skin, tendon and bone, in birds and mammals, the homoiothermic vertebrates, there are two alpha 1 chains and one alpha 2 chain. There are reports of three different chains in chicken bone and cod skin (α_1, α_2 and α_3). The latter contains a higher proportion of basic amino acids (arginine, lysine, histidine) and a lower ratio of the amino acids proline and hydroxyproline to serine and threonine.

The three polypeptide chains form a left-handed triple helix, thereafter they are twisted into a right-handed spiral (fig. 2.1). Each chain has a molecular weight of about 95 000, the whole tropocollagen molecule 290 000.

In view of the large size of the tropocollagen molecule it is not surprising that the primary structure of the individual polypeptide chains is only now being elucidated. The complete amino acid sequence

10

$$\alpha^1 \quad \alpha^2 \alpha^1$$

[a]   [b]   [c]

Fig. 2.1.   Diagram of tropocollagen molecule. (a) Single polypeptide chain coiled in a left-handed helix, (b) coiled chain now wound around axis in a right-handed helix, (c) three chains wound round axis to produce tropocollagen molecule (after Herring).

for the α_1 chain in calf skin collagen was first announced at the end of 1972 by P. P. Fietzek and K. Kuhn.

The α_1 chain is more amenable to amino acid sequencing than the α_2. The alpha chain contains 7 methionyl residues, which can be cleaved by cyanogen bromide, resulting in eight unique peptides, of molecular weights ranging from 1500–24 000 (e.g. α_1–CB8 contains 279 residues). These peptides can in turn be cleaved into two further peptides with hydroxylamine (HA1 and HA2). The latter contains 180 amino acid residues and the sequence of this was worked out by employing enzymes which broke the peptide in different ways. By comparison of the sequence of small peptides (of say 10–20 residues) obtained by breakdown with trypsin, with that for peptides obtained by say chymotrypsin, overlap regions can be detected, and so the complete "jigsaw" pieced together.

The major feature of the amino acid sequence of α_1 calf skin collagen is the central helical portion.

11

Conventionally, residues are numbered from the amino (N-) terminal end of the peptide. Residues 1–16 have been shown to be non-helical; similarly there is a non-helical region at the C-terminus, the final residue being No. 1052.

There does not appear to be any internal homology within the α_1 chain, though one does find a few sequences which are very similar[1]:

Residue 85: Gly–*Met*–Hylys–Gly–His–Arg–Gly–Phe–Ser–Gly–Leu–*Asp*.

Residue 925: Gly–*Ileu*–Hylys–Gly–His–Arg–Gly–Phe–Ser–Gly–Leu–*Glu*.

Proline is hydroxylated to hydroxyproline only when separated from glycine by another amino acid residue (Gly–Pro–Hypro). Every third position along the helical portion of the polypeptide must be occupied by glycine. This allows the larger side chains of the other amino acids to be directed outwards without disrupting the helical packing of the three polypeptide chains. Proline and hydroxyproline make up a further quarter of the residues. The two amino acids hydroxyproline and hydroxylysine are confined to vertebrate connective tissues and hydroxylysine is virtually unique to collagen. In both these cases the hydroxyl group is added on after the protein has been synthesized in a process requiring the presence of vitamin C. This process nevertheless still takes place within the cell. The cross links between the chains appear to be formed extracellularly.

The polypeptide chains of the tropocollagen molecule are stabilized by the formation of intermolecular hydrogen bonds. Similar bonds appear to be present between the tropocollagen molecules. These intermolecular cross-links are especially important, as they confer stability on the collagen fibrils. The number of these bonds increases during the life of the individual.

The tropocollagen molecule is about 280 nm in length and 1·36 nm in diameter. The distribution of charged amino acid residues (positive charge, lysine and arginine; negative aspartic acid and glutamic acid) along the macromolecule is such that the two ends are differently charged, so that the tropocollagen has a head and a tail. The length of the tropocollagen molecule can be determined from the manner in which dissolved native collagen can be reconstituted under different conditions. Under the electron microscope native collagen exhibits a banding with a periodicity of 64 nm (fig. 2.2). This same periodicity can be

[1] Abbreviations of amino acids:
Cys cysteine, Asp aspartic acid, AspN asparagine, Thr threonine, Ser serine, Glu glutamic acid, Pro proline, Hypro hydroxyproline, Gly glycine, Ala alanine, Val valine, Met methionine, Ileu isoleucine, Leu leucine, Tyr tyrosine, Try tryptophan, Phe phenylalanine, Lys lysine, Hylys hydroxylysine, His histidine, Arg arginine.

produced from reconstituted collagen. Other forms include a fibrous form called fibrous long spacing collagen, with a periodicity of 280 nm and a segment long spacing with the segments having a length of 280 nm.

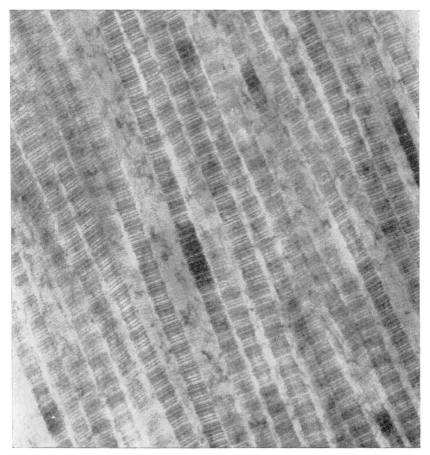

Fig. 2.2. Electron micrograph of collagen fibrils to show 64 nm banding, × 100 000 (Margaret Hackemann).

Only reconstituted collagen with the 64 nm periodicity or demineralized native collagen are able to act as seeding sites for calcium phosphate deposition. The 64 nm periodicity is just under a quarter of the length of the tropocollagen molecule, and this has led to the suggestion that the molecules are arranged side by side in a quarter stagger. This model left a hole of 37·5 nm and an overlap of 26·5 nm, termed the hole and overlap zones (fig. 2.3a).

This model, however, is not accepted by all workers and further models have been produced. Grant and his colleagues have postulated than the tropocollagen molecule is divided into five bonding zones of 26·5 nm in length and four non-bonding zones of 37·5 nm. This model would allow a random stacking, which would automatically give the 64 nm periodicity (fig. 2.3b).

A more revolutionary model has been proposed by Davidovits. In this particular model, the tropocollagen molecule is wound into a spiral with alternating straight and curved portions to give an ovoid disc measuring 64 nm and 50 nm. These discs when aligned together make up the collagen fibrils and give the 64 nm banding (fig. 2.3c). Davidovits believes that the view of tropocollagen as a linear structure accepted by other workers is a type of denatured collagen in which the spiral has unwound.

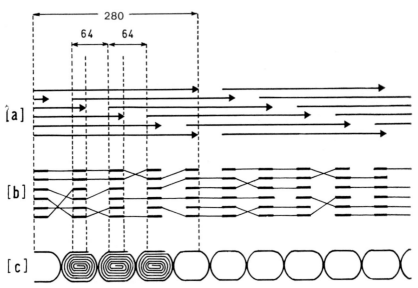

Fig. 2.3.   Packing of tropocollagen molecules according to various theories. (a) Quarter stagger arrangement proposed by Hodge et al., showing overlap and hole zones, (b) random aggregation of bonding (equivalent to overlap) and non-bonding (hole) zones suggested by Grant et al., (c) 'micelle' model of Davidovits (after Herring).

The latest version, the "limiting microfibril hypothesis", has been advanced by A. Veis and his co-workers. This is a modification of the quarter stagger model. The basic unit is a tetramer of four tropocollagen molecules in which the α_2 chains face the centre. The individual tropo-collagen molecules are arranged in a quarter stagger, and the tetramers fit into each other end to end with a 26·5 nm overlap zone and a 37·5 nm hole.

The fossil record throws some light on the various theories of the aggregation of the tropocollagen molecules. In fossil collagens of 50 000 years ago the periodicity has been shown to be about 60 nm. When collagen is extracted from bones and teeth some 20 000 000 years old the periodicity is down to 50 nm. This reduction seems to involve a degree of denaturation that is perfectly regular and would not be expected from the bonding/non-bonding and spiral models. For the present, the tetramer "limiting microfibril hypothesis" would seem to be the most likely.

2.2. Glycosaminoglycans (acid mucopolysaccharides)

Vertebrate hard tissues contain much carbohydrate, a major fraction of which is in the form of chains of amino-monosaccharide units linked by glycosidic bonds, and attached to a protein core. Their physical characteristics are indicated by their designation as the 'snot and slime' polysaccharides. In the past various names have been applied to these important components of the ground substances, such as mucoproteins, mucopolysaccharides and mucins. The current terminology recognizes two major groups, the glycoproteins and the proteoglycans.

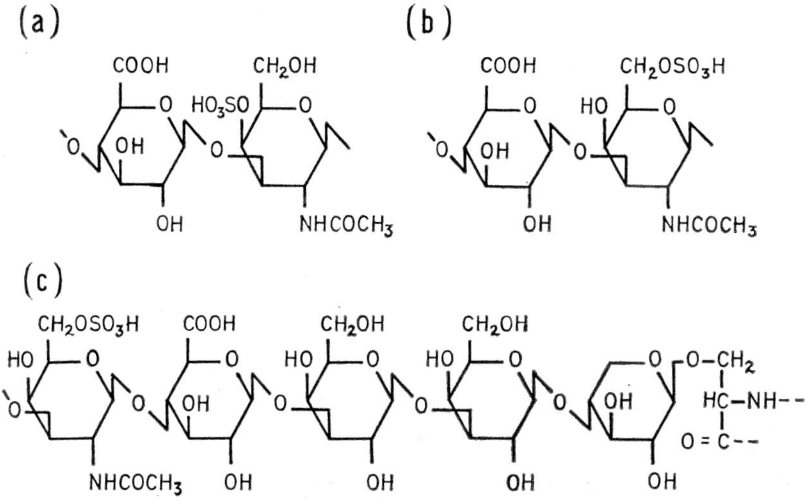

Fig. 2.4. Structural formulae of repeating units of glycosaminoglycans (a) chondroitin-4-sulphate, (b) chondroitin-6-sulphate, (c) linkage region of chondroitin sulphate protein complex, reading from the left: chondroitin sulphate–galactose–galactose–xylose–serine.

15

The glycoproteins comprise short oligosaccharide chains, which contain a wide variety of sugars attached to a protein core. These substances occur in saliva and mucus in general. Of more interest with regard to hard tissues are the proteoglycans, which comprise glycosaminoglycans with long chains of repeating disaccharide units, for example chondroitin sulphates, attached to a protein core.

The chondroitin sulphates are examples of glycosaminoglycans in which the disaccharide units consist of glucuronic acid and acetylgalactosamine. The latter can be sulphated at the 4 position giving chondroitin-4-sulphate or at the 6 position to give chondroitin-6-sulphate (fig. 2.4a and b). The long chains, made up of a repeated disaccharide sequence, are attached to the protein core through a linkage region of neutral sugars, apparently to the serine residues of the protein core. The linkage region consists of galactose–galactose–xylose–serine (fig. 2.4c). The entire macromolecule is believed to consists of some 60 chondroitin sulphate chains, each of a molecular weight of about 50 000, as well as the linkage groups and the protein backbone, so that the total molecular weight is in the region of 4 000 000 (fig. 2.5).

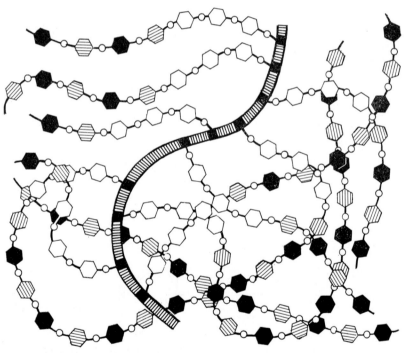

Fig. 2.5. Model of glycosaminoglycan. Protein chain with serine, solid; linkage section (xylose–galactose–galactose), open hexagons; repeating disaccharide units, solid and lined hexagons; oxygen atoms, open circles (after Phelps.)

16

In sharks, only chondroitin-6-sulphate is found in embryonic cartilage. Later in uncalcified cartilage both chondroitin-6-sulphate and chondroitin-4-sulphate occur but the former predominates; in calcified cartilage the chondroitin-4-sulphate predominates. In the bony fishes, there is more chondroitin-4-sulphate but this is absent from the cranial cartilages. With regard to the higher vertebrates, chondroitin-4-sulphate is predominant in cartilage and is the only chondroitin sulphate found in bone. It is evident that chondroitin-4-sulphate is associated with calcification, and according to some authors it may play a major role in bone formation. Whether or not this is so, it is evidently a key macromolecule as regards the calcification of cartilage.

2.3. Keratins

The most readily recognized materials in the animal world are fibrous proteins, the keratins. They form the outer layer of the body, skin, hair, nails, claws, tortoise-shell, horn, feathers, and 'whale-bone' (baleen). Historically, keratin is important in that it was the first protein to be studied with X-rays, and the first one for which Pauling established the helical secondary structure of protein chains.

It was early discovered that there was a regular periodicity of 0·51 nm in mammalian keratins such as wool, but that after being stretched in steam they gave a 0·33 nm spacing. The unstretched condition was named the alpha, the stretched the beta. Subsequently it was discovered that in reptiles and birds the keratins were already in the stretched condition with a periodicity of 0·31 nm. It is now known that birds and reptiles can synthesize both types, whereas the mammals produce only alpha keratin. Keratins are notable for their insolubility which is related to the way in which the molecules are cross-linked with disulphide bonds. Keratin can be broken down into two fractions, one sulphur-rich and made up of globular molecules which act as a cement, and the other with low sulphur and a fibrous structure producing the fibres. Hair consists of microfibrils embedded in an amorphous matrix. Sheep wool has a high sulphur content whereas sheep horn has a low one. The highest comes from raccoon hair, the lowest from rhinoceros horn. The nature of the keratin and its flexibility would seem to be dependent on the sulphur content.

The low-sulphur fibrous fraction forms a left-handed alpha helix which can be further broken down into a helical core and non-helical tails. These tails appear to link the separate units end to end, and also to join the units to the high-sulphur molecules. The helical portions appear to be uniform throughout the mammals, the variation being confined to the non-helical tails (fig. 2.6).

The microfibrils of mammalian keratins show a repeat distance of 20 nm, with a diameter of 7·5 nm. There is evidence that within each microfibril there are eleven protofibrils each with a diameter of 2 nm.

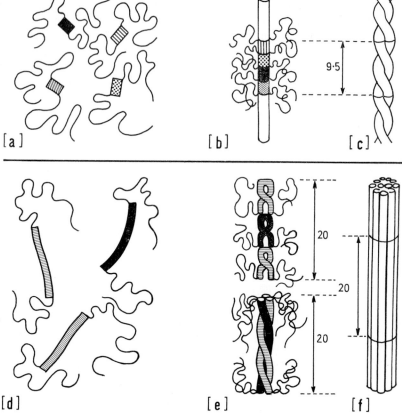

Fig. 2.6. Molecular arrangement of keratin. (a–c) Feather keratin, (a) four different molecules each with central helical portions and non-helical 'tails', (b) aggregation to produce protofibril, (c) coiling of protofibrils to make up microfibril with 9·5 nm periodicity. (d–f) Alpha keratin, (d) two types of molecule in 2 : 1 ratio, (e) two alternative suggestions for aggregation to produce protofibrils, (f) microfibril comprising eleven protofibrils (after Fraser).

Crick has suggested that the protofibrils are made up of two helically coiled polypeptide chains twisted together like a rope. The manner in which these units are put together is still not properly understood and there are several possibilities. If they are arranged in a linear series, a 40 nm periodicity results, whereas a 20 nm one is observed. With an appropriate staggering the correct repeat spacing can be obtained. Other possibilities are that each unit is folded back on itself, or the organization may be of a triple-stranded rope (fig. 2.6 d–f).

In marked contrast, the 'feather' keratins of reptiles and birds have microfibrils with a 3·5 nm diameter and with a repeat distance along the microfibrils of 9·5 nm. A pair of protofibrils coil round each other to produce the microfibril (fig. 2.6). Each protofibril unit comprises four molecules which together make a turn through 9·5 nm. The zig-zag beta configuration allows the two twisted strands to mesh.

It seems clear that in evolutionary terms the mammalian keratins represent an advance on the reptilian and bird forms. Presumably it should be possible to suggest the sequence of changes from one form to the other, but for the present this cannot be done.

Another feature that may be relevant in this discussion is the distribution of calcified keratins. F. G. E. Pautard and his colleagues have identified crystallites of the mineral apatite, calcium phosphate, in rhinoceros horn, platypus hair, lion whiskers, goose feather calamus, and human finger nails. In a study of the hairs on the soles of the feet of polar bears, they found that the apatite crystals were aligned along the keratin fibrils. Claws and bird beaks have zones of appreciable calcification. But the most highly mineralized keratinous tissue is baleen, the huge epidermal plates hanging from the roof of the mouths of the whale-bone whales.

There is a close correlation between the degree of calcification of keratin and the amount of the amino acid proline. Calcified keratins contain large amounts of proline in contrast to hair, for example. Apart from mammoth hair and the extinct South American ground sloth, there are few records of fossil keratins. The earliest record is of tortoise-shell from the Oligocene (40 million years ago) of the Isle of Wight, in Britain. As with the tortoise-shell of living forms, the proline content is high, and it can be inferred that the fossil form was calcified.

The evolutionary significance of calcified keratins and the increase in sulphur content is still to be determined. It is evident that the water-proofing role of a keratinous epidermis in the first fully terrestrial verte-brates, the reptiles, eventually evolved into the insulating function of hair and feathers. The specialized hair in the mouth of the whale-bone whales, with its high degree of calcification, may represent a neo-formation—a redevelopment of an earlier calcified condition.

2.4. *Enamel protein*

In developing partially mineralized enamel, there is 15–20 per cent by weight of organic matter; in mature enamel it is reduced to 0·35 per cent. Very little is known about either foetal or adult enamel protein. It is not known whether there is a single protein or several. From the amino acid composition it is possible to say what enamel protein is not. The small amount of glycine, which makes up a third of the collagen molecule, the minute amount of alanine and large amount of glutamic acid, militate against the identity of enamel protein with collagen (fig. 2.7).

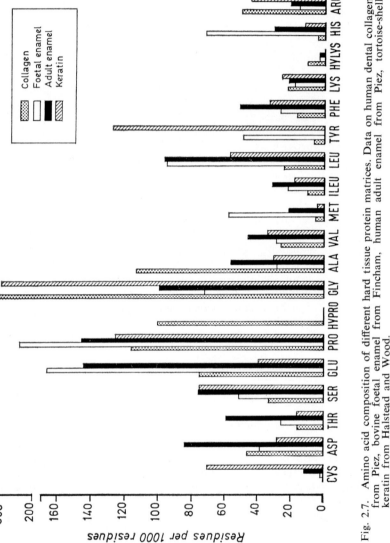

Fig. 2.7. Amino acid composition of different hard tissue protein matrices. Data on human dental collagen from Piez, bovine foetal enamel from Fincham, human adult enamel from Piez, tortoise-shell keratin from Halstead and Wood.

At one time it was proposed that enamel protein belonged to the eukeratin family of proteins, but the presence of histidine, lysine and arginine in the ratio of $3:1:1$ contrasts with the eukeratins, where the ratio was $1:4:12$. Cystine is present in very minute amounts in contrast to the situation in keratin (fig. 2.7). In view of these compositional differences the identification with keratins can be excluded.

The large amounts of proline have led to the suggestion that the foetal protein is in the form of an amorphous gel rather than as fibres. In fact, X-ray studies indicate the presence of unoriented protein molecules; in any case there are few signs of fibrous material being present. Perhaps the most unusual aspect of enamel protein is the remarkable change in composition between the organic matrix of developing and that of mature enamel. During development the amount of proline and histidine is enormously reduced, whereas the proportional amount of glycine is increased (fig. 2.7).

Enamel protein is considered to be thixotropic, i.e. it has the ability to flow under pressure, but no one has yet been able to trace the course of events that occurs during the maturation of enamel. What is clear is that most of the enamel protein disappears, but it is not certain how it does so or where it goes. The relationship of the protein matrix to the crystallites of hydroxyapatite is also unknown.

CHAPTER 3

mechanism of calcification

The calcified tissues, and particularly bone, exist in a dynamic equilibrium with blood; this means that a knowledge of the components of blood plasma is an essential prerequisite for the understanding of the process of calcification. Many of the more important pieces of evidence were obtained from experiments *in vitro*. One of the first problems was to understand why bone mineral, hydroxyapatite, can form in the body at the concentrations of calcium and phosphate found in the body, when *in vitro* it would require three times as much for the mineral to precipitate.

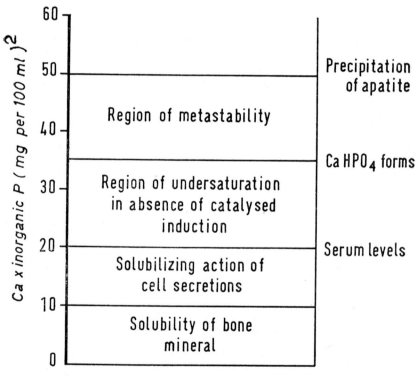

Fig. 3.1. Calcium and phosphate products to precipitate apatite in relation to their levels in serum and bone mineral solubility product (after Simkiss).

If bone mineral, hydroxyapatite, from bone is shaken with physiological saline, it dissolves, and a near equilibrium is obtained when the product of the concentrations of calcium and phosphate ions reaches 10 (mg per 100 ml)^2. At this point the mineral breaks down into ions— this is termed the dissociation product. If on the other hand calcium and phosphate ions are added to physiological saline, the mineral calcium phosphate will be precipitated only when a product of between 35 and 50 (mg per 100 ml)^2 is reached (fig. 3.1).

Now, the naturally occurring product in plasma is 20 (mg per 100 ml)^2 and this means that plasma is undersaturated as regards the precipitation of mineral. For many years it has been a mystery how apatite was brought out of solution at such low ionic concentrations. On the other hand, plasma is supersaturated with regard to the dissociation of bone mineral, and it is therefore difficult to see how skeletal tissues can be resorbed.

The paradox was resolved by W. F. Neuman and M. W. Neuman who put forward the thesis that bone mineral did not precipitate, but instead developed by a process of crystallization. For precipitation it is necessary to postulate 18 molecules coming together simultaneously, and this requires a considerable concentration. In contrast, crystallization implies the growth of crystal lattices by the gradual addition of the appropriate ions. This can be accomplished at much lower concentrations, but for its initiation there needs to be a template on which the crystals can grow.

It is believed that the structure of collagen with its reactive sites is such a template. This is the epitaxy theory of calcification. Strong support for the idea that collagen acted as a nucleator or template for the deposition of calcium phosphate came when Neuman and Neuman showed that when collagen was added to the solutions, apatite came out of solution (crystallized) at concentrations of 20 (mg per 100 ml)^2, that is at the level found in serum. It was thus clear that the presence of collagen was an important factor in the process of calcification, that it was a major activator (fig. 3.2).

These experimental studies seemed to have finally answered the question of how tissues are able to calcify. Unfortunately this answer brought in its wake an even greater problem. It was no longer a question about why tissues became calcified but rather why tissues did not calcify. Collagen is one of the main constituents of connective tissues and is one of the most widespread of all proteins, comprising a third of the total protein in the body. Yet in general it does not calcify, except under pathological conditions. It was suggested by several authors that there must be some kind of inhibitor that prevented calcification taking place. Unfortunately no one was able to find it.

This inhibition hypothesis received strong support when an ultra-filtrate of plasma was added to the solutions. When this was done calcium phosphate came out of solution at 28 instead of 20 (mg per 100 ml)[2]. Clearly the nucleating activity of collagen was being inhibited by some substance. After further research H. Fleisch and his co-workers came to the conclusion that pyrophosphates were the main inhibiting agents. The fact that these substances occurred in the urine in much greater concentrations than in plasma explained the fact that the urine could be supersaturated with calcium and phosphate ions without vesicular calculi forming. These phosphates seem to act by adsorbing on to the collagen fibrils, and in this way preventing the apatite from seeding. Pyrophosphates, by blocking the seeding sites, are considered to be crystal poisons.

Adenosine triphosphate (ATP) is known to give rise to adenosine monophosphate (AMP) and pyrophosphate ions in the presence of apatite. The existence of bone mineral by itself would seem to induce the production of its own inhibitor.

One of the first attempts to unravel the calcification story was made in 1923 by H. Robison, who discovered that at all the sites of calcification the enzyme alkaline phosphatase was present. He contended that this enzyme was responsible for hydrolysing organic phosphates to increase the free phosphate ions so that the solubility product would be exceeded in the extracellular finds thus resulting in the precipitation of calcium phosphate. Since that time, it has been shown that alkaline phosphatase does not play an active role in the process of calcification. However, it is always present, and it does break down pyrophosphates. It is now generally accepted that *in vivo* pyrophosphatase destroys the crystal poisons and in this way allows calcification to proceed. From the earlier experiments, in which alkaline phosphatase was added to the collagen plus plasma ultrafiltrate, calcification proceeded in much the same way as when only collagen was added. This experiment proved that alkaline phosphatase prevented the inhibitor from functioning.

There is today a fairly well established body of data regarding the details of the process of calcification. But this is not the whole story. Collagen has a key position in the process just described, but it is now recognized that other organic macromolecules can also act as seeding or nucleating sites; the unrelated keratins, and the glycosaminoglycans, to say nothing of the mysterious enamel protein. Clearly the entire process does not depend on collagen alone.

More recently, evidence has come to light that the calcium phosphate is not at first in a crystalline form, which indicates that the crystallization concept may not be entirely valid. Indeed, some authors have even gone so far as to suggest that crystalline hydroxyapatite is an artefact caused by the preparation techniques. It has been demonstrated that the initial apatite granules are associated with the collagens and that these initial deposits have a 64 nm spacing. Recently it has been sug-

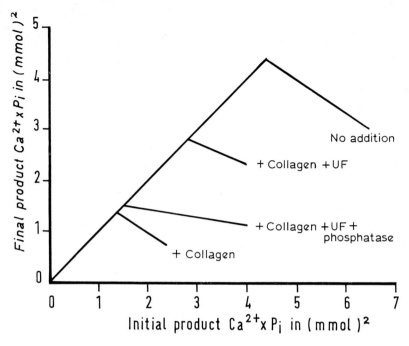

Fig. 3.2. Results of experiment showing influence of collagen, plasma ultra-filtrate and alkaline phosphatase on the formation of calcium phosphate crystals (after Fleisch and Neuman).

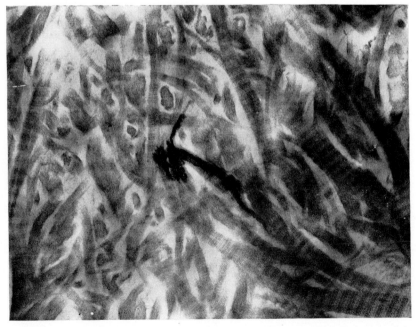

Fig. 3.3. Electron micrograph showing apatite crystallites developing in association with collagen fibrils, ×50 000 (A. F. Hayward).

gested that these observations were artefacts. Subsequent development of these nuclei is for the crystallites to grow so that the C axes are parallel to the collagen fibrils (fig. 3.3). The initial spacing has led to the suggestions that the formation of bone mineral begins in the 'holes' of the tropocollagen collagen fibril formed by the quarter-stagger packing of the tropocollagen molecules. Certain authors have pointed out that in many instances the apatite crystallites grow without any apparent orientation to the alignment of the collagen fibrils.

Where the crystallites are obviously associated with the fibrils they seem to form a series of rings around fibrils. There is still some controversy over this, as it is contended that the crystallites also develop inside the collagen fibrils. The fact that the collagenous matrix can survive hundreds of millions of years preserved in fossil bones and teeth seems to suggest that the collagen is enclosed by mineral rather than being merely an embedding material.

Major aspects of the mechanism of calcification are now well established. Perhaps the most critical area is how the dynamic equilibrium is maintained between the skeleton and the blood, what M. R. Urist terms the 'bone–body fluid continuum'. This is regulated by hormones and is discussed in the following chapter.

CHAPTER 4

mineral homoiostasis

The concentration of ionic calcium in the blood is maintained at an appropriate level, but this has to be closely regulated in living organisms. As with all biological systems, this regulation is effected by means of a negative feedback system. The response to a stimulus results in an effect on the original stimulus; this is a feedback and such control systems are termed servo systems. Any variation from the norm produced by the stimulus will result in an effect that will reduce this variation by modifying the stimulus accordingly; hence the term negative feedback.

If blood plasma calcium levels fall excessively, the nerves and muscles will discharge automatically and the muscles will go into spontaneous spasms—a condition known as tetany. The calcium level is increased by the action of a hormone, the parathyroid hormone, which stimulates bone resorption by osteocytes and osteoclasts and the extraction of calcium from the urine. Vitamin D, which, because it is synthesized in the skin, is considered by some authors to be a hormone, also assists in the maintenance of calcium levels by increasing calcium uptake by the cells of the intestine. When the correct level has been reached the supply of hormone is cut off and an equilibrium is attained. The variation in the production of amounts of parathyroid hormone was thought to be responsible for the maintenance of calcium plasma levels. As parathyroid hormone remains in the blood stream, it would be expected that the levels would overshoot; furthermore it should not be possible to reverse the system rapidly, but it was observed that it is.

In 1962 D. H. Copp proposed that another hormone must exist which acts as an antagonist to parathyroid hormone. This postulated hormone, which he named calcitonin, lowered the level of calcium and inhibited bone resorption. The evidence he presented consisted of a series of elegant experiments. Initially he perfused the parathyroids of dogs. Solutions were introduced into the arteries supplying the parathyroids and collected from the veins leaving the glands. In this way the organ was isolated and given different types of information so that its reactions could be monitored. A high calcium perfusion led to a rapid fall in plasma calcium, a low calcium perfusion produced a rapid rise in plasma calcium.

After a low calcium (hypocalcaemic) perfusion the parathyroids were removed and the plasma calcium level continued to rise. When a high calcium (hypercalcaemic) perfusion was substituted for a parathyroid-ectomy there was a fall instead of the expected continued rise. Copp concluded from this that there must be a factor which was responsible for lowering the plasma calcium levels. He then made a further experiment to confirm the existence of this factor. He found that if the parathyroids were removed after a parathyroid perfusion there would be a continued rise of calcium, indicating a release from an inhibiting factor. This was the hypercalcaemic overshoot expected, if calcium levels were controlled simply by the parathyroid hormone alone. Only after 36 hours would an equilibrium be established.

The hormone calcitonin was later discovered to be secreted by the C-cells of the thyroid gland. In mammals these cells are derived from the ultimobranchial glands that become incorporated into the thyroid gland. In fish, amphibia, reptiles and birds these glands remain as discrete entities and in all these groups are now known to produce calcitonin. The presence of calcitonin in sharks reveals that this hor-mone is not primarily concerned with inhibiting bone resorption and osteoclast activity, at least not in evolutionary terms. Rather it must originally have been the major hormone regulating calcium levels in the blood. The main problem for marine organisms is to prevent the calcium levels becoming too high. Parathyroid hormone is absent from sharks and their allies. Vitamin D is present in the bony fishes and may play

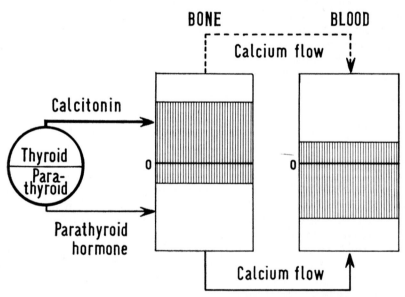

Fig. 4.1. Diagram illustrating control loop regulating balance of calcium in bone and blood by calcitonin and parathyroid hormone (after Rasmussen and Pechet).

an analogous role to parathyroid hormone. It is ironic that the most primitive of these regulating substances should have been the last to be discovered.

In the mammals, calcitonin and parathyroid hormone work in concert to maintain the calcium levels (fig. 4.1). With an increase in the level of plasma calcium there is a rise in the secretion of calcitonin and at the same time the amounts of parathyroid hormone are decreased. If the calcium level becomes too low then parathyroid hormone is secreted in increased amounts. With these two hormones it becomes possible to control the level of plasma calcium effectively and to prevent the problems that would be consequent on overshoot.

Since Copp's original proposal that there was a new hormone, calcitonin, the primary structure of the molecule has been determined— it comprises a sequence of 32 amino acids. The sequence differs slightly from one species to another. Advances have since been made in understanding the exact mechanism by which calcitonin acts to inhibit bone resorption. However, to understand this it is necessary to consider the action of parathyroid hormone at the same time. When bone has been formed, it is surrounded by a thin layer of liquid covered by resting bone-forming cells, osteoblasts. Enclosed bone cells, osteocytes, are also bathed in tissue fluid in the lacunae of bone. The osteocytes and resting osteoblasts regulate the flow of ions between the blood plasma and the tissue fluid. Parathyroid hormone affects the osteocytes, which leads to the breakdown of the collagen matrix and the concomitant release of calcium to the blood. Remodelling of bone, which involves the removal of some of it, is accomplished by special bone-destroying cells, osteoclasts, but the normal process of maintaining mineral homoiostasis involves only the osteocytes. With the breakdown of collagen, hydroxyproline is excreted in the urine. The amount of hydroxyproline in the urine is in fact used as a rough measure of the amount of bone resorption taking place. Similarly there are generally increased amounts of calcium present in the urine. With injections of calcitonin the amounts of hydroxyproline and calcium in the urine are markedly decreased. The activity of the osteoclasts is increased with parathyroid hormone and reduced by doses of calcitonin.

The action of parathyroid hormone is two-fold: on arrival at the cell membrane it stimulates the uptake of calcium by the cell from the surrounding tissue fluid and activates the enzyme adenyl cyclase. This enzyme acts on adenosine triphosphate to produce cyclic adenosine monophosphate which blocks the movement of calcium from the cytoplasm to the mitochondria (see Chapter 5). At the same time calcium continues to flow into the cytoplasm from both the extracellular tissue fluid and from the mitochondria, so that there is an appreciable accumulation of calcium in the cytoplasm. This stimulates the cells concerned to begin the resorption of the surrounding bone by the breakdown of the collagenous matrix. The action of calcitonin is to activate the cell's calcium pump, which transports calcium from the cytoplasm across the cell membrane and into the extracellular tissue fluid. This action switches off the process initiated by parathyroid hormone, and may furthermore be an important stage in bone formation.

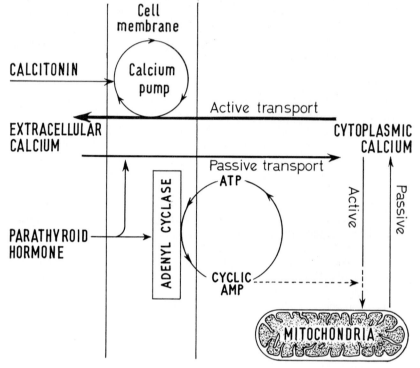

Fig. 4.2. Action of calcitonin and parathyroid hormone on the cell (after Rasmussen and Pechet).

Within the space of ten years, the existence of a new hormone has been postulated and it has been identified, extracted and now synthesized. Its mode of action also has been worked out.

Although we now have a clearer idea as to how the bone–body fluid continuum is maintained, there still remains the question how bone is formed in a situation where the calcium level in the blood is comparatively low. The answer to this seems to reside in the activities of the organelle called the mitochondrion.

CHAPTER 5
role of mitochondria

Calcification is controlled by many factors, but there would still seem to be a need for some means of concentrating calcium and phosphate ions to enable the process to proceed. During the 1960's evidence accumulated pointing to the mitochondria of the cell being involved. The first piece of evidence was the recognition that isolated mitochondria accumulated large amounts of calcium ions. This was surprising because it was believed that calcium ions were toxic to mitochondria as they inhibited the main activity of these organelles.

In 1948 A. L. Lehninger discovered that the enzymes involved in the citric acid and respiratory cycles were located in the mitochondrion.

The mitochondrion is popularly known as the power-house of the cell since it is here that energy is released, from the oxidation of foodstuffs, in the citric acid or Krebs cycle. The main function is the conversion of adenosine diphosphate and inorganic phosphate to adenosine triphosphate, which is then exported into the cell, where it is split into adenosine diphosphate and phosphate with the release of energy. These two products are then returned to the mitochondrion where the process begins again. The adenosine triphosphate may in some instance be broken down to adenosine monophosphate and the energy-rich pyrophosphate (fig. 5.1). The enzyme pyrophosphatase hydrolysed the pyrophosphate to two phosphate ions. Adenosine monophosphate is converted back to the triphosphate in two stages. First it reacts with adenosine triphosphate to give two molecules of diphosphate, and then the normal formation of adenosine triphosphate can take place.

The building up of adenosine triphosphate (oxidative phosphorylation) and calcium ion accumulation were found to be two alternative processes, both being energy dependent.

Furthermore, the uptake of calcium ions was accompanied by an increased intake of phosphate ions. The intra-mitochondrial calcium/phosphate molar ratio was 1·7, remarkably close to that of the mineral hydroxyapatite, which is 1·65. Only in the presence of phosphates are calcium ions released from the inner mitochondrial membrane. Only in the presence of adenosine triphosphate or diphosphate are deposits of electron-dense granules formed, giving rise to the 'massive loading' of the mitochondria. This process will only take place in the presence of a respiratory substrate. Without adenosine diphosphate the mitochondria simply swell, and only accumulate limited amounts of calcium ions.

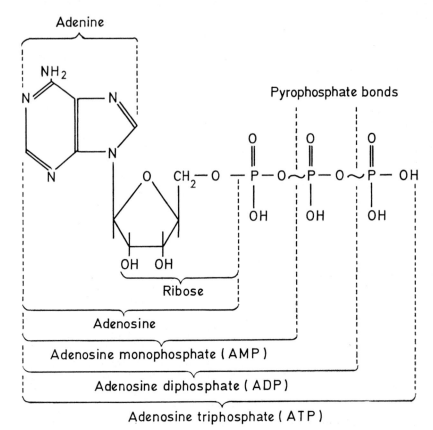

Fig. 5.1. Structure of adenosine triphosphate (ATP). The terminal phosphate groups are attached by energy-rich bonds. Hydrolysis releases this energy with the formation of either adenosine diphosphate (ADP) or mono-phosphate (AMP) (after Clowes).

Without phosphate anions the electron-dense granules do not form.

The observations were initially made on isolated mitochondria, but the electron-dense granules were also seen in mitochondria in rat liver.

The next stage was the recognition that calcium ions were preferentially taken up by the mitochondria as part of the process of egg-shell formation in birds. The large freshwater protozoan *Spirostomum* is known to produce a type of internal 'skeleton' of calcium phosphate and the mitochondria contain amorphous calcium phosphate. Pautard suggested that the transition from the amorphous to the crystalline state might occur in the mitochondria.

All these observations led I. M. Shapiro and J. S. Greenspan in 1969 to propose as a working hypothesis that biological mineralization was under strict cellular control and that the mitochondria were the cellular organelles responsible for concentrating the calcium and phosphate ions. They postulated that the ions thus concentrated were then transferred to the mineralizing sites of calcification. The extracellular levels of calcium and phosphate were raised, and in the presence of an appropriate matrix apatite crystallites were seeded.

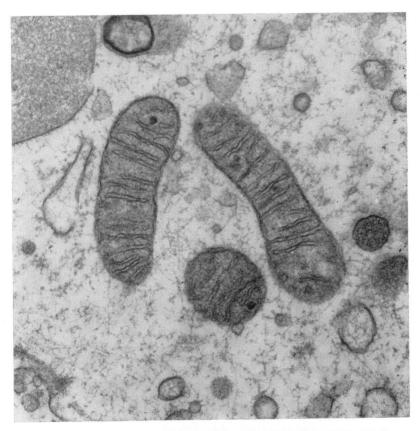

Fig. 5.2.   Electron micrograph of mitochondria × 35 000 (A. F. Hayward).

Subsequent researches have tended to confirm the view that the mitochondria are responsible for the early stages of calcification. It has now been discovered that the mitochondria in calcifying cartilage are loaded with calcium phosphate granules (fig. 5.2). These granules are entirely amorphous. From the X-ray diffraction studies that have been carried out, there is no sign of any hydroxyapatite pattern. Recently it has been noted that there are two phases of calcium phosphate formation in bone:

33

a non-diffracting amorphous one which seems to be tricalcium phosphate, and the crystalline, approximating to hydroxyapatite. It now seems clear that tricalcium phosphate represents the first stage in the formation of crystalline calcium phosphate. Previously it was believed that bone crystals were formed on collagen fibrils direct from calcium and phosphate ions in the extra-cellular fluid. This view was explicitly accepted by Shapiro and Greenspan, although their hypothesis implied that the mitochondria might be doing more than concentrating ions.

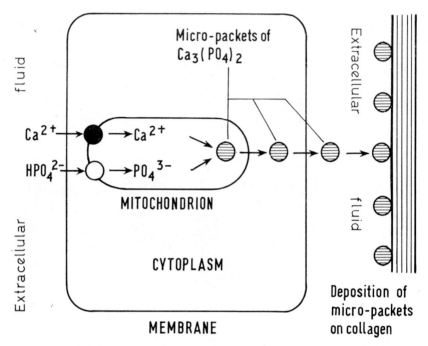

Fig. 5.3. Hypothesis to show probable role of mitochondrion in biological mineralization (after Lehninger).

The conversion of tricalcium phosphate to apatite takes place rapidly *in vitro* but in living systems is not so easy, because at the pH of plasma (7·0) tricalcium phosphate is more soluble than hydroxyapatite. Hence, when blood plasma and interstitial fluid are supersaturated with respect to hydroxyapatite, they will be undersaturated in respect of tricalcium phosphate. It is thus necessary to have a mechanism for concentrating the calcium and phosphate ions to such an extent that the solubility product of amorphous tricalcium phosphate is exceeded.

Lehninger developed the Shapiro–Greenspan hypothesis starting from the initial situation in which calcium and phosphate ions are present in the intercellular fluid but where they are undersaturated. These ions are concentrated by the mitochondria, and the result of this localized concentration is that tricalcium phosphate comes out of solution. Micropackets of this mineral are believed to be stabilized by some as yet unidentified inhibitory factor, probably pyrophosphate, and then transferred from the mitochondria, through the plasma membrane of the cells, to attach by epitaxis to the appropriate sites on the collagen fibrils, where they become part of the amorphous fraction of the bone mineral (fig. 5.3).

There are still a large number of unknown factors in Lehninger's theory, but at the moment it can be suggested that the mitochondria are responsible for the concentration of calcium and phosphate ions, and that these are not transferred to the sites of mineralization in the ionic condition but as tricalcium phosphate.

It is now established that the mitochondria are more than the phosphorylating power plants they were first thought to be. They do indeed retain this as their primary function but on occasions this can be overridden. The same energy-dependent system that is responsible for oxidative phosphorylation can also be utilized for transporting calcium ions against gradients. It has been suggested that the formation of calcium phosphate granules in mitochondria may be a waste disposal system that only comes into operation when the normal metabolic activity of the cell is in abeyance. The processes of calcium accumulation and oxidative phosphorylation appear to be complementary. The accumulation and subsequent expulsion of calcium ions or calcium phosphate were originally an excretory process that has since become transformed into an integral part of the calcification process.

The fact that mitochondria are capable of these two contrasting functions may be related to the recently proposed suggestion that they originated from independent organisms. It has long been known that mitochondria exhibit a degree of independent behaviour in the cell, in that they are able to reproduce, even when the cell itself is not undergoing mitosis. Mitochondria are unique among the organelles of the cell in that they possess their own DNA, which is different from that located in the nucleus. Furthermore all the structures necessary for protein synthesis are located within the mitochondria, they have transfer RNA, messenger RNA and ribosomes. The length of the double-stranded DNA of the mitochondrion is such that only some 5000 amino acids can be coded. The mitochondrial DNA molecules are circular, and in this respect are comparable to those known from bacteria and blue-green algae. It has been suggested that the mitochondria may be related to these primitive organisms, which at an early stage in the evolution of life on this planet entered the larger cells of other organisms. A comparable situation is known in the giant amoeba *Pelomyxa palustris* which does not possess mitochondria but instead houses symbiotic bacteria which play an active part in the cell cycle.

If this hypothesis is true, the mitochondria, from being initially parasites, developed during the course of evolution into endosymbionts. Their presence would, in all likelihood, have conferred an advantage by virtue of the extra means of creating energy. When extracted they are no longer capable of surviving independently, so they have to rely on the host cell for many of their key metabolic processes. At the same time they play an important role in the life of the cell. The mitochondria are passed from generation to the next only via the ovum—they take no part in meiosis.

The excretory role of mitochondria—to be expected in a one-time independent organism—is of particular significance, when it comes to considering the origins of vertebrate hard tissues.

CHAPTER 6

origin of vertebrate calcified tissues

The origin of bone or bone-like tissues in the vertebrates has exercised the minds of many scientists over many decades. Since bone is confined to the vertebrates and the success of this sub-phylum can in a large measure be attributed to the possession of an internal bony skeleton, it is not surprising that this question has been the subject of much speculation.

The first vertebrates of which there is any record were the ostracoderms of the Ordovician and Silurian. They were jawless microphagous feeders and are associated in classification with the modern cyclostomes, from which they differed in possessing a bony armour, by which we know of their existence. The earliest bony tissue thus formed an exoskeleton and this in itself gives a clue to the origin of bone. In the first instance its function was clearly not that of an internal support. C. F. A. Pantin and others taught in the 1920's that bone was, at least in part, an excretory product; calcium phosphate sometimes precipitates in human urine. Many of the earlier theories were based on the assumption that the development of an exoskeleton was related to the conquest of fresh waters by the early vertebrates or pre-vertebrates.

By far the most widely known theory was proposed by A. S. Romer, who contended that, as the name armour suggests, the primary role of the exoskeleton was for protection, in this case against eurypterids, which were giant freshwater scorpions. Yet many of the young ostracoderms were unarmoured and they managed to survive. Although the arthropods with their huge pincers looked fierce, it is hard to believe that they could have been a serious danger to the mud-grubbing ostracoderms. In any event the very first ostracoderms were marine-living, and probably never came in contact with the eurypterids from one generation to the next.

Homer Smith, best noted for his work on the kidney, believed that ostracoderm armour acted as a kind of waterproofing necessitated by the invasion of freshwaters. Without such a covering, water would enter the tissues by osmosis leading to oedema and death. As Romer pointed out, the structure of the armour was inside out, as the outer portions comprised a spongy open meshwork of bone with a basal lamellar layer; the entire organization was singularly inappropriate as a waterproofing agent. Further, the bony covering had already been acquired in the sea before the ostracoderms colonized fresh waters.

N. J. Berrill also associated the acquisition of a bony exoskeleton with the taking up of life in fresh waters. He considered that the bone was an excretory product—a means of getting rid of the excess phosphates the animals would have met on entering rivers. There is no evidence for such excess phosphates in fresh waters, except for our own effluents— hardly likely to have been present some 500 million years ago. A comparable argument was put forward by T. S. Westoll suggesting that it was to get rid of the excess calcium found in fresh water as the early vertebrates would have been unable to cope with all their intake of calcium. Westoll seems to have hit on the most likely explanation but for quite the wrong reasons.

About ten years ago the discussion on the origin of bone returned to the premise that this event took place in marine conditions. At that time Pautard suggested that bone was initially laid down as a phosphate store. As there was an abundance of calcium in the sea, to store phosphate in the form of calcium phosphate seemed a reasonable postulate. Pautard believed that the shortage of phosphate in ancient seas would lead to the success of any organism capable of acquiring its own phosphate store. Unfortunately there is no reason to suggest that the amount of available phosphates in the sea at the time of the origin of the vertebrates was basically any different from what it is today. The amount of phosphates in the seas is in fact extremely small, but in spite of this the oceans support a considerable volume of living matter. It is not a question of there not being enough phosphate but rather of its distribution.

There is a seasonal phosphate cycle in the sea which begins in the spring when the waters are stirred by gales and phosphate ions become available for plants. They are not in a form assimilable by animals, and are taken up by the phytoplankton, which multiplies in the upper zones of the sea until all the available phosphates are used up. This is the optimum time for the supply of phosphates for animals feeding on the phytoplankton. As the surface waters are fairly calm the salts are not replenished from below, and eventually, as there is a continual loss from the upper layers, the plankton begins to die off towards the end of the summer, and so do the animals. The phosphates, as a result of the disintegration of the organic matter, are again in the form of free ions. Although during the winter there is a maximum amount of free phosphate in the sea, this is not a form that can be utilized by animals. In these circumstances any organism that can acquire a private phosphate store during the period of summer abundance will have a considerable advantage during the winter dearth.

The above theories were current until Shapiro and Greenspan formulated their hypothesis on the role of mitochondria in biological mineralization. With the advent of this work it became necessary to re-examine the different theories on the origin of bone. It is now possible to propose a comprehensive postulate which in fact incorporates all the major theories that have previously been advanced. It is simply that bone has played quite different roles at different times in its evolutionary history.

As Homer Smith surmised, the origin of a vertebrate calcified tissue was connected with problems of osmosis, although not quite in the way he imagined. In the sea there is an excess of calcium ions which will enter the organism from the environment. In order to maintain a reasonable degree of homoiostasis, there has to be some kind of calcium pump to excrete the excess calcium. Of necessity this process will involve the expenditure of energy. As noted in the previous section, the mitochondrion is the cellular organelle involved in this process. The original calcified deposit is likely to have been a by-product of calcium excretion as Westoll suggested. It seems reasonable to suppose that during the seasonal abundance of assimilable phosphates the calcium would have formed calcium triphosphate and would have been transferred extracellularly to the skin, which is an ideal site for the deposition of waste material, a site moreover notable for the concentration of collagen. Calcium phosphate deposited in the skin would have been effectively removed from the system.

Fortuitously, such deposits would have represented a valuable phosphate store during the period of seasonal dearth. Possession of such stores would have conferred important advantages on the animals concerned.

The oldest known calcified tissue includes calcified cartilage, aspidin (a type of bone-like tissue), dentine and bone. All these tissues, to be discussed in the next section, were part of a covering of plates forming the armour of the ostracoderms. There seems no reason to deny these tissues a protective role in the life of these particular animals. Even if we do not accept that freshwater scorpions were involved, there were other organisms that could have acted as effective predators. It seems difficult to accept that these highly organized tissues evolved initially as a protective device, but once calcium phosphate deposits were already in existence there was a possibility of their modification for such purposes as protection, a role which they retain in other animals.

From serving as an exoskeleton, the next stage was for such tissues to replace the internal supporting cartilaginous skeleton to produce a bony endoskeleton. We do not know what could have prompted this important change. Perhaps light will be thrown on this in the future when more of the evolution of the collagen molecule is understood. It is difficult to understand the advantage of a bony internal skeleton among fully aquatic organisms. On land there is an obvious advantage in the possession of a bony internal skeleton, but in water a cartilaginous skeleton would seem to be perfectly adequate; the sharks and their allies are still remarkably successful. Curiously the ancestors both of the sharks and of the bony fish originally had a bony exoskeleton in conjunction with a cartilaginous endoskeleton. The sharks lost the bony exoskeleton—all that remains is a shagreen of placoid scales; the bony fish acquired a bony endoskeleton and also reduced the exoskeleton. A possible clue to the change might be found in the ostracoderms, in one group of which bone from the dermal armour gradually invaded and replaced the presumably cartilaginous tissue of the head region. Be that as it may, when the vertebrates ventured on to land, the bony internal skeleton was one of their most important assets and so it remains to this day.

Once the vertebrates left the water the skeleton took on a further role, that of a calcium store. Aspects of this have already been discussed in Chapter 4 and will be dealt with when the problems that a laying hen encounters in producing egg-shells are discussed.

Bone has played a variety of roles during its long evolutionary history and apart from the two initial ones of calcium excretion and phosphate store, it still retains the other three—protection, internal support and calcium store.

The amount of bone mineral required for the chemical store and protection is much less than the amount present. Most of the skeleton is concerned with support and if this function is removed by either immobilization or the state of weightlessness, the body takes active steps to reduce the amount of the skeleton. In osteoporosis, a disease of old age, up to 50 per cent of the bone mineral of trabeculae can be lost. This is a major problem in space travel, as astronauts lose appreciable amounts of their skeletons during flights. During the Gemini IV flight, which lasted four days, one astronaut lost between 1 and 12 per cent of his bones; after the eight day Gemini V flight, a 20 per cent loss was recorded for some bones. Exercises for the astronauts in the Gemini VII trip reduced the losses somewhat but the problem remains a serious one that is still to be overcome.

the cells of hard tissues

7.1.1. *Protein synthesizing and secreting cells*

The areas where hard tissues are formed are generally characterized by an increase in the degree of vascularity. The formation of the different hard tissues involves the activity of cells, that are derived from the mesoderm. Their main activities are the synthesis and secretion of the organic matrix on which calcification takes place. For this reason they are known as protein synthesizing and secreting cells. They have a

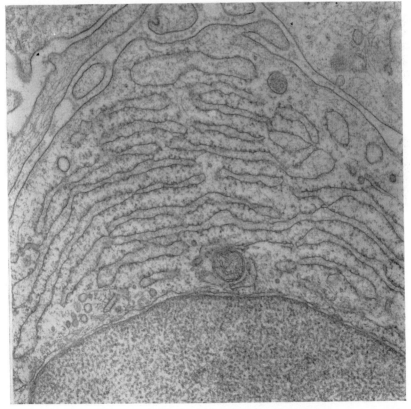

Fig. 7.1. Electron micrograph of rough endoplasmic reticulum of a protein synthesizing and secreting cell, ×25 000 (Margaret Hackemann).

high content of RNA, a prominent rough endoplasmic reticulum and Golgi apparatus, and often numerous mitochondria and secretory vesicles. Protein synthesis takes place in the ribosomes of the rough endoplasmic reticulum (fig. 7.1); the protein is then secreted from the lumen of the endoplasmic reticulum to the Golgi apparatus, and thence to the secretory vesicles.

Fibres and a ground substance are then passed out of the cell, although the passage of fibres of collagen through the Golgi apparatus has been questioned. The tropocollagen molecules are believed to be assembled into fibrils extracellularly.

Enamel—the most highly mineralized of all vertebrate hard tissues—is also produced by protein synthesizing and secreting cells but this time they are derived from the ectoderm. The rough endoplasmic reticulum is very prominent and there is a high RNA content. In contrast with cells that form bone and dentine, the ameloblasts (the enamel secreting

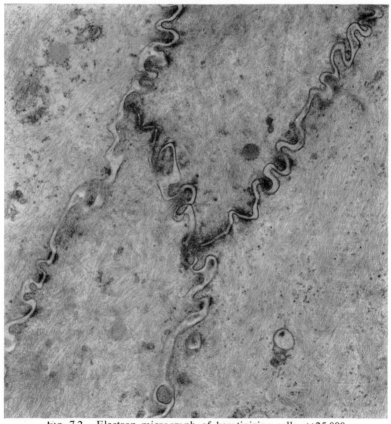

Fig. 7.2.   Electron micrograph of keratinizing cells, ×25 000
(Margaret Hackemann).

cells) do not show the presence of the enzyme alkaline phosphatase. This enzyme, which is associated with calcification of the organic matrix, occurs in the cells of the stratum intermedium which lie adjacent to the ameloblasts.

7.1.2. *Protein synthesizing and retaining cells*

In marked contrast to these cells are those in which the protein synthesized is retained within the cell. This type is termed the protein synthesizing and retaining cell. Here there is little rough endoplasmic reticulum, but instead there are numerous isolated ribosome rosettes. The protein that is synthesized remains in the cell until the cytoplasm becomes laden (fig. 7.2). All the keratinous tissues are formed on this principle, and the calcification that takes place is similarly intracellular.

Protein synthesis is the first and perhaps the most critical aspect of the genesis of hard tissue but is only a part of the whole. In many tissues there are, as well as cells producing the organic matrix, others enclosed within the calcified material and there are further specialized cells that have the ability to resorb hard tissues. The best known of these three types of cell are those from bone, called respectively osteoblasts, osteocytes and osteoclasts, which are briefly discussed below.

7.2. *Osteoblasts*

In the mesenchyme are numerous generalized cells that have the potentiality to develop into more specialized cells; because of this ability they are termed pluripotent. The pluripotent stem cells develop along along two major lines, the haemopoietic that gives rise to blood cells, and the skeletogenic which gives rise to the cells forming fibrous tissue (fibroblasts), bone (osteoblasts) and cartilage (chondroblasts). The pluripotent mesenchyme cells which develop along the osteogenic or bone line may become preosteoblasts or preosteoclasts. Fibroblasts may either develop into osteoblasts or remain fibrogenic, depending on the conditions in which they find themselves.

The osteoblasts generally form a pavement covering newly formed bone and mediate in the transfer of calcium from the blood to the matrix. This matrix is collagenous. The culturing of osteoblasts *in vitro* confirms that they have the ability to synthesize copious amount of collagen. Osteoblasts do not divide, so that there is a constant recruitment from the mesenchyme progenitors.

Osteoblasts tend to be columnar with the nucleus furthest away from the developing bone surface. The rough endoplasmic reticulum and mitochondria are concentrated at the opposite pole from the nucleus (fig. 7.3), which is rich in RNA. The surface of the osteoblast in contact with the newly formed organic matrix of bone is thrown into numerous projections which penetrate the matrix. It is believed that these cyto-

43

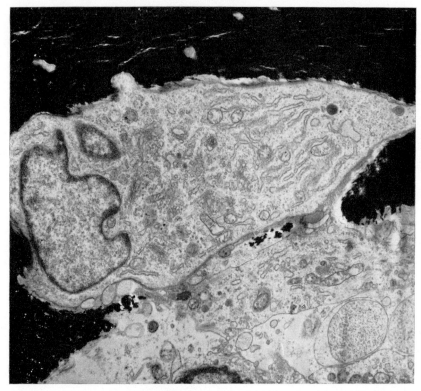

Fig. 7.3.   Electron micrograph of osteoblast, partially enclosed in calcified matrix ×7000 (J. Appleton).

plasmic extensions may be concerned with organizing the collagenous framework, as well as the calcification of the matrix. The preosteoblast is concerned with dividing to maintain the population of osteoblasts. The amount of RNA in the nucleus and cytoplasm of the preosteo-blasts is comparable to that of the nucleus of osteoblasts but the amount in the cytoplasm of the latter is about three times as much again, con-firming that a large amount of protein synthesis is being conducted by the osteoblasts.

7.3. *Osteocytes*

As bone formation proceeds the osteoblasts in closest proximity to the newly formed bone matrix become walled in. Once this has occurred they are known as osteocytes. At first they are indistinguishable from osteoblasts, but subsequently they undergo a number of changes. They are extremely difficult to study as they are enclosed in bone tissue and are consequently difficult of access (fig. 7.4).

44

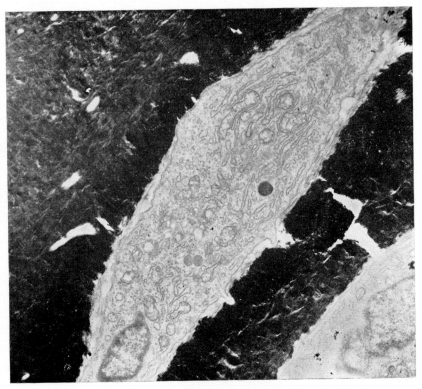

Fig. 7.4. Electron micrograph of osteocyte × 7000 (J. Appleton).

In mature osteocytes there is a gap between the cell and the wall of the lacuna but there is still argument as to whether this is an artifact produced during the preparation of the sections. Recently a number of authors have contended that there is a protein–polysaccharide lining between the cell membrane and the wall of the lacuna. It has been known for a decade that when fossil bone is macerated one is left with a collection of isolated 'osteocytes' which are composed of a chemically inert polysaccharide (fig. 7.5). The recognition of the existence of such a lining in the living tissues has come much later than its discovery in fossils.

Deeply situated osteocytes are notable for the poor development of their organelles. The endoplasmic reticulum is sparse, as is to be expected in a resting cell. The mitochondria contain dense granules.

It has been observed that in the space between the cell membrane and the wall of the lacuna there may be loose collagen fibrils as well as an amorphous matrix. The latter is generally identified as a proteoglycan (protein-polysaccharide). A. Boyde, from his scanning electron micro-

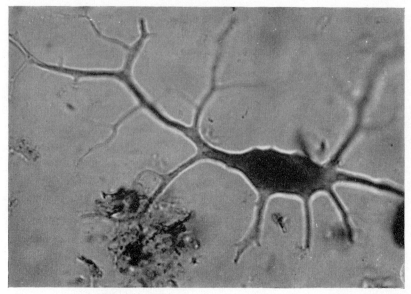

Fig. 7.5. Micrograph of lining of osteocyte lacuna and canaliculae from Carboniferous fish bone, × 1800 (photo J. R. Mercer).

scope studies, has shown that the lining of the osteocyte lacuna in the resting state has a comparatively smooth textured collagenous lining (fig. 7.6).

The role of the osteocyte is still not properly understood. As discussed in Chapter 4, it seems likely that the osteocytes play an important role in the maintenance of mineral homoiostasis. It has also been suggested that they are responsible for the synthesis of proteoglycans. Cell processes pass into the canaliculae but there are no organelles present in these extensions.

7.4. Osteoclasts

Cells are needed for the formation and maintenance of hard tissue, and it is equally important that there should be cells capable of destroying it. Such cells are the osteoclasts, which are large, multinucleated cells, some containing several hundred nuclei. Osteoclasts originate from the fusion of precursor cells. Signs of their activity are seen in the scallopped resorption pits, the lacunae of Howship. Evidence of resorption in the earliest bone-like tissue in the fossil record was recorded for the first time only ten years ago.

One of the most characteristic features of osteoclasts is the 'brush border' which is in direct contact with the bone surface. N. M. Hancox and B. Boothroyd have produced some cine films of cultured osteoclasts

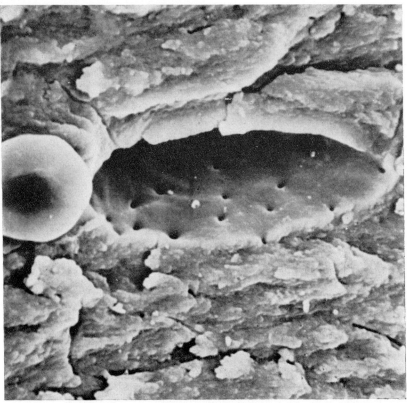

Fig. 7.6. Scanning electron micrograph of lacuna of resting osteocyte and surrounding bone, scale given by erythrocyte (photo supplied by Dr A. Boyde, University College London).

that reveal vigorous activity on the part of the brush border. In electron micrographs this area is seen to be a region of pinocytosis with finger-like processes of the cytoplasm. These invaginations give way to vacuoles in the cytoplasm, and within these fragments of collagen fibrils and apatite crystallites can be found (fig. 7.7). It seems as if the collagen fibrils are being stripped of their apatite. The endoplasmic reticulum is not conspicuous, but mitochondria are numerous. The action of parathyroid hormone increases the amount of messenger RNA from the nucleus to the cytoplasm. Presumably this is concerned with the synthesis of enzymes. There are many lysosomes, the organelles concerned with breaking down substances. Indeed the digestion of the constituents of the bone is the concern of the lysosomes, but it is not known how the bone is dissolved at the brush border. However, it seems that this is one of the few known examples of extracellular lysosomal

47

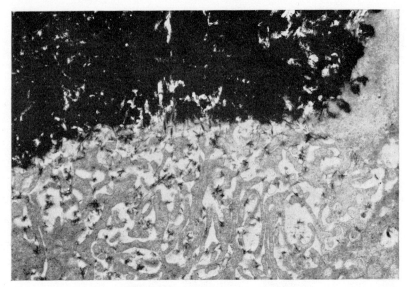

Fig. 7.7. Electron micrograph of osteoclast brush-border, showing removal and incorporation of crystallites from bone ×8500 (photo supplied by Dr B. Boothroyd, Liverpool University).

activity. It is not known how the crystals are detached from the collagenous matrix, whether this is accomplished by lysosomal enzymes breaking down the matrix or more directly.

In spite of the problems still remaining, great strides have been made in the last few years in our understanding of the way in which the osteoclast functions.

PART II

THE TISSUES

cartilage

8.1. *Cartilage types*

Of all the skeletal tissues to be discussed, cartilage is unique in that it is capable of growth by internal expansion. Basically cartilage is composed of cells, the chondrocytes, embedded in a gel-like matrix of chondroitin sulphate with some collagen fibrils. Growth of cartilage is by two processes. To begin with, chondroblasts are formed from fibroblast-like cells in the perichondrium, and as they produce the organic matrix around themselves the perichondrium retreats. This type of growth is by apposition and is characteristic of nearly all hard tissues. Unlike other hard tissue cells, the enclosed chondrocytes do not lose the ability to divide, and are capable of undergoing mitosis; more important, they are able to produce further matrix. Typically, chondrocytes occur in clones of two, three, four or more cells. The formation of additional intercellular matrix in this way allows the cartilaginous structure to increase in size without in any way altering adjacent or attached structures such as muscles, nerves and blood vessels. This property of cartilage led Romer to suggest in a famous essay written 30 years ago that cartilage was an embryonic adaptation; cartilage acted as a skeletal template and was able to grow without there being a need to alter the relationship of the associated blood vessels and muscles. Clearly this is one of the main advantages of cartilage as a skeletal tissue, but there is no need to postulate that this material is specially developed for the needs of the embryo. Such a view seems reasonable when consideration is given only to modern land vertebrates, but it is hardly necessary to invoke such a theory when considering such aquatic animals as sharks.

8.1.1. *Hyaline cartilage*

The type of cartilage that has been most intensively studied is hyaline cartilage, so called because of its somewhat glassy appearance (Latin; *hyalinus,* glass). Its development can be considered in two stages. First there is a zone of proliferation of chondroblasts, which are produced by the mesenchyme cells of the perichondrium. As the cells produce the intercellular matrix they gradually separate. When this separation is well marked and the cells have acquired their rounded outlines, the zone of maturation is reached. The chondrocytes are situated in lacunae and are capable of dividing to give cell nests or clones. The chondrocytes can further separate by the secretion of still further matrix (fig. 8.1).

The articular surfaces of bone are covered by a specialized type of cartilage. In this articular cartilage, the superficial surface in contact with the synovial fluid of the joint comprises small cells which are

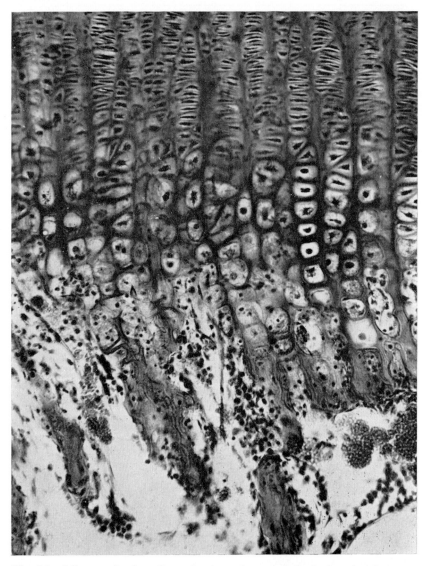

Fig. 8.1. Micrograph of cartilage showing columns of developing chondrocytes, hypertrophy, together with region of osteoblasts and bone formation in lower part of photograph ×245 (photo J. R. Mercer).

flattened with their axes parallel to the joint surface; beneath this layer the cells are larger and are arranged in columns at right angles to the surface. In sections of developing joints there is a zone of proliferating cartilage cells beneath the articular surface, but the articular cartilage does not seem to be replenished from this proliferating zone. Collagen fibrils in the matrix run at right angles to the surface between the vertical columns but at the surface itself become tangential.

The zone of maturation grades into a zone in which the cells hypertrophy (swell up) and the intervening matrix becomes impregnated with crystallites of apatite.

8.1.2. Elastic cartilage

There is a type of cartilage in which the chondroitin sulphate matrix has numerous fibres of elastin scattered throughout it. This is termed elastic cartilage and is found in regions of the body where structures while needing to be stiff must also be elastic, for example in mammalian ears and epiglottis.

8.1.3. Fibrocartilage

Where cartilage requires an increase in tensile strength, there is an appreciable increase in the amount of collagen fibrils in the matrix, hence the term fibrocartilage. Most of the tissue consists of bundles of collagen, between which are rows of mature chondrocytes. Fibrocartilage is developed at the point where tendons and ligaments are inserted into either bone or cartilage. This tissue is also the main material in fibrous joints such as the pubic and mandibular symphyses.

Intervertebral discs also contain fibrocartilage. The articular surfaces of the vertebral centra are covered in hyaline cartilage, and the surfaces of two adjacent vertebrae are joined by a kind of capsule of fibrocartilage which encloses a central space containing a pulp fluid, the nucleus pulposus. The nucleus pulposus is a surviving remnant of the notochord. The fibrocartilage is known as the annulus fibrosus (fig. 8.2). This arrangement gives the vertebral column a degree of resilience. Loss of water from the nucleus pulposus in old age causes the intervertebral discs to shrink and the spine to shorten.

During the healing of a bone fracture, an enclosing callus of "fibrocartilage" is developed. It is not unlike hyaline cartilage except that the amount of collagen present is much greater. The cartilage becomes calcified and is replaced by bone which forms a bony union of the fracture. This topic is discussed in more detail in Chapter 21.

8.2. Endochrondal ossification

The manner in which the cartilaginous skeleton of the embryo is transformed into that of the bony one of the adult has led to a number

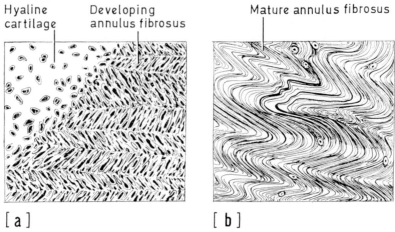

Hyaline cartilage | Developing annulus fibrosus | Mature annulus fibrosus

[a]   [b]

Fig. 8.2. (*a*) Developing fibrocartilage of intervertebral disc, (*b*) mature fibro-cartilage of intervertebral disc.

of theories regarding the evolutionary relationship of bone to cartilage. John Hunter, the eighteenth century anatomist, put the matter succinctly as follows—'bone is not the original skeleton in any animal, but only the adult; for in the first formation of any animal which afterwards is to have bone, the skeleton is either membrane or cartilage, which is changed for bone, but not into bone.'

If the hyaline cartilage is examined from the zone of proliferation through that of maturation to the region where bone is being laid down, there is an intervening zone of calcified cartilage. In this zone the chondrocytes become hypertrophic. They swell up and the intervening matrix between them becomes reduced in volume and becomes calcified. In epiphyseal plates the hypertrophic chondrocytes form vertical columns between which are struts of matrix. The changes that take place in these cells are not well understood but they seem to herald the onset of calcification of the matrix. The chondrocytes die and as they disintegrate there is some breakdown of the calcified matrix (fig. 8.3).

The death and breakdown of the chondrocytes induce an increased vascularization and the invasion of the area by osteogenic cells. Osteo-blasts line up along the surface of the vertical struts of calcified cartilage. Bone is then laid down on these struts which thus become incorporated within its structure.

8.3. *Calcification of cartilage*

Until fairly recently it was assumed that calcification in cartilage took place on the collagen fraction of the organic matrix, but it is now estab-lished that collagen plays no role in this process as far as cartilage is

54

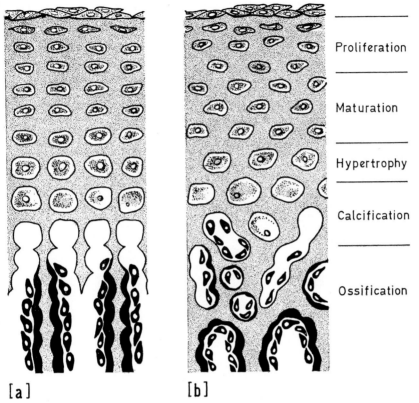

Proliferation

Maturation

Hypertrophy

Calcification

Ossification

[a]   [b]

Fig. 8.3. Endochondral ossification, (a) in long bone, (b) in mandible (after Ham).

concerned. Initial calcification seemed to be in spherulitic aggregations of apatite crystallites in the regions of the matrix furthest removed from the cells themselves in extracellular organelles the matrix vesicles. By means of special decalcification techniques, two workers, E. Bonnucci and J. Appleton, independently discovered that the apatite crystallites were formed in association with organic 'crystal ghosts'. These appeared to be structures now recognized as matrix vesicles within which calcification is initiated. The available evidence indicates that the organic material is a proteoglycan, chondroitin-4-sulphate. This work on the calcification of cartilage is important as it establishes that calcification can take place on a variety of organic matrices. It also suggests that a spherulitic type of calcification may well be a primitive form of mineralization.

Among the very first vertebrates in the fossil record from the Ordovician period, 500 million years ago, there are pieces of calcified cartilage. These belong to the heterostracan ostracoderm *Eriptychius* and are positioned just beneath the bony armour of the head region. The plates of calcified cartilage are smooth externally, but the limiting shell is formed by coalesced spherulites which produce a wave-like front. Within this delimiting shell there are spherical spherulites (fig. 8.4). This type of spherulitic calcification of cartilage is known within the dermal armour of one of the last representatives of another group of ostracoderms, the cephalaspids, distantly related to the lampreys, and in the antiarch placoderms, one of the first groups of jawed vertebrates. It is also known in some of the living primitive sharks.

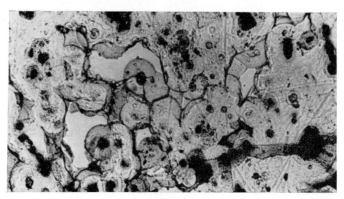

Fig. 8.4. Micrograph of spherulitic calcified cartilage in the dermal armour of the Ordovician heterostracan, *Eriptychius*, × 135 (photo J. R. Mercer).

In the heterostracan ostracoderms, which are considered to be ancestral to the jawed vertebrates, calcified cartilage is known only in the earliest representative, *Eriptychius*. It has not been recorded in any later heterostracan species. Its appearance in the last examples of certain other groups such as cephalaspids and antiarchs does not necessarily mean it is an advanced feature. Frequently at the end of an evolutionary line, there is a reappearance of primitive conditions.

If the sharks are considered, it becomes clear that the spherulitic type of calcification is the primitive state. In the more advanced forms the calcification of cartilage becomes much more highly organized. In all cases cartilage is never completely calcified, rather there are zones of calcification. This always allows nutrients to pass through the un-calcified areas to maintain the vitality of the tissue. The patterns of calcification can be used to distinguish different genera of sharks from one another.

The prismatic calcification of the different elasmobranchs is of two main types, a radial and a circular arrangement (fig. 8.5).

Among the sharks calcified cartilage seems to be perfectly viable. Even in the most highly calcified parts of the cartilage, the cells remain healthy and there is no tendency for them to disintegrate. It remains something of a mystery why calcification in the cartilage of the higher vertebrates seems to be associated with the breakdown of the tissue and the death of the chondrocytes. Presumably this is, in evolutionary terms, a fairly recent development. Among the sharks, calcified cartilage is the major tissue of the skeleton.

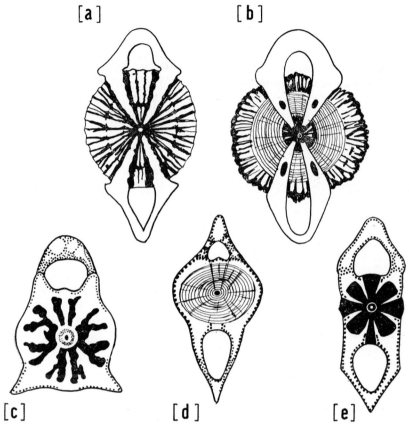

Fig. 8.5. Prismatic calcified cartilage in vertebrae of elasmobranchs (sharks and rays). (a) *Carcharodon*, (b) *Cetorhinus*, (c) *Orectolobus*, (d) *Squatina*, (e) *Rhinobatis* (after Ridewood).

57

CHAPTER 9
aspidin

9.1. *Structure of aspidin*

The main part of the dermal armour of the first vertebrates, the heterostracan ostracoderms, is a bone-like tissue for which W. Gross in 1930 coined the non-commital term aspidin. The basal layer comprises thin lamellae penetrated by vertical ascending canals which lead to an open spongy meshwork, with the macroscopic appearance of cancellous bone. This middle layer of the armour consists of a three-dimensional scaffolding of calcified trabeculae of varying dimensions (fig. 9.1). In some groups the trabeculae form vertical walls enclosing polygonal spaces, but even in these more specialized forms the cancellae break up into the normal spongy texture at the edges of the plates, and also towards the uppermost layers.

The most notable feature of aspidin is that it does not have normal osteocyte lacunae and canaliculi enclosed within it. This has resulted in much controversy regarding the relationship of this tissue to bone. There are two schools of thought, one which contends that aspidin is a secondary derivative of bone, and the other which believes that aspidin is a primitive tissue from which bone can be derived. In fact when aspidin from different geological periods is studied, differences in structure can be recognized and it becomes possible to outline the evolutionary changes that must have occurred.

The earliest and apparently most simple type of aspidin is a three-dimensional scaffolding of narrow trabeculae in which no structures can be discerned. In this case it is reasonable to infer the existence of aspidinoblasts, which lined up much as osteoblasts do when they produce the organic matrix for the initial spicules of bone, from which the trabeculae of cancellous bone are developed. Subsequent to this stage the aspidinoblasts would have secreted the organic matrix from one surface only. The cells responsible for the formation of the matrix of aspidin did not become incorporated into the tissue; rather, as aspidin was produced the cells retreated. In certain specimens it is evident that there are two types of aspidin which parallel the situation in bone. First there are the initial trabeculae which form the basic scaffolding, and then successive layers of aspidin are deposited by apposition on the trabeculae. This lamellar aspidin can thus be compared directly with lamellar bone which is similarly laid down by apposition on initial trabeculae. The general organization of aspidin is indistinguishable

58

Fig. 9.1. Block diagram of heterostracan dermal armour to show basal layer of lamellar aspidin, middle layer of spongy aspidin and superficial dentine tubercles (from L. B. Halstead, *The Pattern of Vertebrate Evolution.* Oliver & Boyd).

from the spongy or cancellous bone of other vertebrates. This tissue is an acellular type of bone.

This is by no means the whole story, because there are structures enclosed within the aspidin of later heterostracans, that have been the subject of controversy during the last decade. In most examples of aspidin, there are spindle-shaped spaces arranged in a somewhat irregular manner or at right angles to the lamellae (fig. 9.2). The generally accepted interpretation was that these structures represented the former position of collagen fibres. At the same time it was stated that the acellularity of aspidin was secondary, and that this tissue had evolved from a more normal bone tissue with included bone cells. In 1963 a new interpretation of the nature of aspidin was proposed. It was pointed out that there were two types of structure within the tissue, a fine-calibre tubule produced by the cell processes of retreating aspidinoblasts, and a larger-calibre one which was formed by the enclosure of simple spindle-shaped aspidinocytes in the collagenous matrix they had secreted. The view that these structures were spaces left by collagen fibrils was rejected and it was suggested that aspidin was a likely evolutionary precursor of bone.

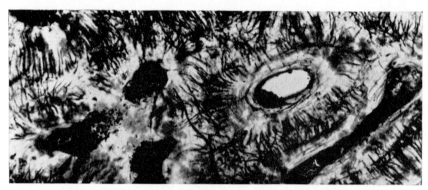

Fig. 9.2. Micrograph of advanced aspidin to show aspidinocyte spaces × 75 (photo J. R. Mercer).

In the Ordovician heterostracan genus *Astraspis,* the aspidin is penetrated by numerous fine tubules at right angles to the surface of apposition. As with osteoblasts, there must have been fine cell processes invading the matrix, which as well as organizing the collagenous frame-work, may have played an important role in calcification. As more matrix was secreted the aspidinoblasts would have retreated, leaving fine tubules penetrating the hard tissue. This interpretation has led some authors to describe this tissue as a type of dentine, in which tubules are similarly formed by the retreating cells maintaining cell processes extend-ing into the tissue.

The spindle-shaped spaces in the aspidin, now identified as aspidino-cyte spaces, are arranged in a random fashion and are always separate from each other. Furthermore, with the apposition of lamellae, these spaces extend towards the vascular spaces in the tissue (that is, at right angles to the lamellae) and gradually taper off as the cells were unable to keep pace with the further material being deposited on the trabeculae. This phenomenon is exactly comparable to what happens to the enclosed osteocytes in the initial trabeculae of bone and in the specialized type of bone, cementum, found around the roots of teeth.

As well as these two types of structure in aspidin, the fine tubules and spindle-shaped spaces, there is a further one that is particularly prominent

60

in the basal portions of the tissue. Parallel tubes are seen running obliquely through the tissue or at right angles to the margins and these can readily be interpreted as the former position of bundles of collagen—Sharpey's fibres that served to anchor the bony plates into the dermis.

There are three possible ways of interpreting the different spaces within the hard tissue aspidin; all three structures may be considered as having housed collagen fibrils, or bundles of collagen and cell processes, or bundles of collagen, cell processes and cell bodies.

9.2. *Evolution of aspidin*

If aspidin from successive geological ages is examined, it becomes possible to trace a number of gradual evolutionary changes over a period lasting 150 million years. From an initially acellular condition, there arose a situation where the aspidinoblasts became incorporated within the matrix of the initial trabeculae. In the most advanced forms, there was a further trapping of cells in the lamellae laid down against the trabeculae. In this regard the organization of the aspidinocytes is comparable to the organization seen in bone. Aspidin demonstrates a possible way in which bone could have arisen from a primitively acellular tissue.

The situation is a little more complex than this, since among the first examples of aspidin, there are types with fine tubules running at right angles to the surface of deposition of the matrix. The formation of such tubules by the cell processes of retreating cells is similar to the situation found in dentine, but it seems likely that there is more than one process per cell, unlike the situation in dentine. Nevertheless, this type of aspidin may suggest a possible origin for dentine. As far as the evolution of aspidin is concerned, this fine tubule development arose in two separate later evolutionary lines of heterostracan ostracoderms culminating in the two genera *Obruchevia* and *Ganosteus*. Both these animals can be traced from genera which possessed aspidin with large aspidinocyte spaces, hence this fine dentine-like structure must have arisen independently from the similar looking material seen in the Ordovician.

There is a further way in which the evolution of aspidin can be traced; it is possible to outline the gradual changes in the organization of the organic matrix. This is achieved by indirect means. When sections are examined with a polarizing microscope, it is possible to work out the organization and orientation of the mineral component of aspidin or bone or dentine. In the latter two tissues, the apatite crystallites are aligned along the collagen fibrils, and by inference it seems likely that the same applied in aspidin. It is known that the organic matrix of aspidin was also collagenous and therefore this seems a reasonable assumption to make. From a study of the orientation of the mineral, it becomes possible to infer indirectly the organization of the organic matrix.

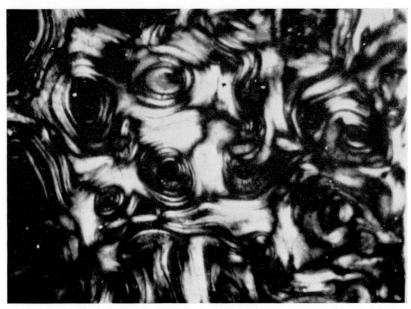

Fig. 9.3. Micrograph of advanced aspidin under polarized light to show narrow black and white banding, ×75 (photo J. R. Mercer).

In the early heterostracans the trabeculae, and the initial scaffolding of some of the advanced types of aspidin, suggest that the aspidino-blasts secreted successive parallel layers of collagen in the same way that dentine-forming cells (the odontoblasts) are known to do. As one traces the development of aspidin, first a broad black and white banding becomes evident. In the most advanced examples, this had developed into a series of narrow black and white bands which are almost impossible to distinguish from those seen in bone (fig. 9.3). This observation suggests that the organization of the organic matrix in aspidin gradually evolved from a dentine-like condition to one characteristic of true bone.

9.3. *Remodelling of aspidin*

Although from a purely structural point of view it can be demonstrated that aspidin had close affinities to bone, it has been stressed by a number of workers that it was fundamentally different physiologically because it was unable to remodel. In this respect it was considered to be more akin to dentine. Gross, many years ago, claimed that aspidin was remodelled, because bony plates of the armour fused and the characteristic spongy margins were replaced by large cancellae. He was, however, unable to find direct evidence of this remodelling. Recently, incontrovertible evidence of resorption in aspidin has been recorded.

The scallopped margins of lacunae of Howship cutting across lamellae of aspidin established the fact that aspidin-destroying cells, or aspidino-clasts, must have existed. It is now generally accepted that aspidin is capable of remodelling, albeit to only a slight degree. In this respect it heralds the condition that becomes extremely well developed in bone.

All the recent work on aspidin points to this tissue being a primitive type of bone. The reason for retaining the name aspidin is that this tissue shows a gradual evolution from a substance that is close to dentine to one that is more obviously allied to bone. To classify it with either would be to obscure the relationship with the other. Aspidin is particularly important in evolutionary studies, as it is one of the very few documented examples of the gradual evolution of a hard tissue.

CHAPTER 10

bone

10.1. Cancellous and compact bone

The basic type of bone formation begins with a clump of osteoblasts embedded in a mat of coarse interlacing fibres of collagen. The cells are in contact by means of their cell processes. When these initial spicules are calcified, the cell bodies and cell processes become enclosed in lacunae and canaliculi respectively. This coarse-fibred bone is termed woven bone; the cells are more numerous than in ordinary bone and the lacunae are more spherical.

Subsequently osteoblasts cover the surface of these spicules, and secrete further collagenous matrix on them in which the fibrils are arranged in a single direction. The osteoblasts become incorporated in the newly formed bone tissue and, in the mammals particularly, become

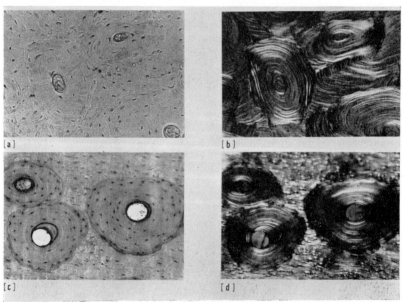

Fig. 10.1. Micrographs of transverse sections of osteones of compact bone. (a) Ground section showing osteocyte lacunae and canaliculae, osteones delimited by marked cement line, (b) same view under polarized light, (c) decalcified section, (d) same view under polarized light × 90 (photos J. R. Mercer).

somewhat almond-shaped and aligned parallel to the surface of apposition. With the constant recruitment of new osteoblasts, the bone surfaces remain covered by cells. The formation of bone, apart from the initial spicules, is by the apposition of new layers or lamellae. This type of bone is termed lamellar. The growth on the spicules produces fine beams or trabeculae and these form an anastomosing three-dimensional scaffolding (see Chapter 19). This spongy tissue is known as cancellous or trabecular bone.

With the addition of numerous lamellae lining the spaces in the tissue, the spaces become gradually smaller and smaller so that the bone looks solid to the naked eye. When this condition has been reached the bone is described as compact or cortical. Concentric rings of lamellae, termed osteones, are typical of most bones (fig. 10.1). Although the terms cancellous and compact are subjective, in practice there is little difficulty in distinguishing them from one another. The outer parts of bones tend to be compact, the inner cancellous.

In many texts reference will be found to membrane bone and cartilage bone. These two types are indistinguishable and the terms lead to some confusion as they imply that bone can be derived from either membrane or cartilage, which is not so. These terms refer to the environment in which bone first develops. In membrane bone the osteoblasts develop in areas where there are many fibroblasts and collagen fibrils, which give a membranous appearance in the embryo. In cartilage bone there is a cartilaginous template of the individual bone, which is replaced by bone. It is not changed into bone. The only justification for using these terms is that they locate the original site of the particular bone. For example, bones that developed from an original cartilaginous endoskeleton have a different evolutionary history from those that were originally of dermal origin. This is not, unfortunately, always a reliable guide. For example, the collar bone, the clavicle, is a cartilage bone in mammals and birds but it can be traced back to part of the dermal armour of prehistoric fish and amphibians.

10.2. *Growth of long bones*

All limb bones begin as cartilage templates. The cartilage is replaced with the exception of the hyaline cartilage of the articular surfaces. In birds and mammals the bones are made up of three parts—the shaft, or diaphysis, and the two ends or epiphyses. The epiphyseal discs are active growth zones which remain cartilaginous until the adult size is reached, whereupon these discs themselves ossify. The advantage of this system is that the articular relations are not involved with the process of growth of the individual bones (fig. 10.2).

The cells surrounding the cartilaginous precursor are pluripotent mesenchyme cells. Fibroblasts and osteoblasts develop from this region, known as the perichondrium. Once osteoblasts derived from the perichondrium begin to produce bone, it is normal to term the surrounding membrane the periosteum. The beginning of the process of ossification is in the central portion of the shaft (the diaphysis), where the chondrocytes hypertrophy and there is a general breakdown of the material. Into this region there come osteoblasts accompanied by capillaries and pluripotent cells which may develop into either osteogenic or haemopoietic cells. Struts of bone are laid down in this central area—the centre of ossification. The formation of bone proceeds from this centre toward the two extremities. While osteoblasts replace the hypertrophied chondrocytes within the diaphysis, osteoblasts from the periosteum lay down successive lamellae of bone to form the compact cortex of the shaft.

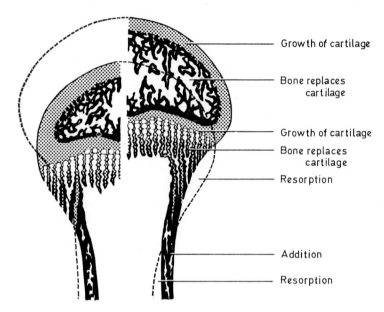

Fig. 10.2. Diagram to illustrate growth of long bone at cartilaginous epiphyseal plate but also involving resorption and deposition in different parts of the diaphysis (after Ham).

Within the centres of the cartilaginous epiphyses further ossification centres develop. Whereas in the case of the diaphysis the osteogenic cells are derived from periosteal buds, in the epiphyses the invasion by capillaries and osteogenic cells is only possible via the route of the blood vessels supplying the epiphyseal cartilage.

The formation of bone in the epiphyses and the diaphysis spreads until there is only a pad of cartilage separating these units. This zone, however, is one of active proliferation of cartilage. As the chondro-

66

cytes mature, they swell up or hypertrophy; there is breakdown accompanied by calcification and the cartilage is replaced by bone. Further cartilage is produced from the region of proliferation to keep pace with the amount that breaks down. In this way an active zone of growth is maintained while the shaft of the bone is increased in length.

While all this bone formation is proceeding within, the osteoblasts produced from the periosteum are laying down consecutive layers of bone. In this way the girth of the bone is similarly increased. If this were all, the shaft would become gradually thicker and thicker. In fact the actual thickness of the compact cortex of the shaft is maintained at a fairly constant amount; this is due to the activity of osteoclasts which ensure that the overall shape of the bone is retained (fig. 10.2). Much of the original cancellous bone of the main part of the diaphysis is also resorbed. The space in the bone houses the developing haemopoietic tissue, the bone marrow, which is concerned with the production of blood cells. Finally, when the bone has reached its definitive size the cartilage of the epiphyseal plate ceases to continue the generation of new cartilage and is replaced by bone. In this way the epiphyses finally fuse to the diaphysis.

10.3. *Laminar and Haversian bone*

The nature of the compact bone of the cortex of limb bones has been studied in many types of vertebrate. This part of the bone has to take up the main forces acting on the bone, in particular the weight of the animal. It also requires an effective blood supply to maintain its vitality. The major bones of large mammals and the extinct dinosaurs have a type of bone that has been named laminar by J. D. Currey, who made a study of its vascularization and its strength.

Laminar bone is made up of a series of laminae between which there are anastomosing networks of blood vessels. This has led one German worker to describe this type of bone as 'wire-netting bone', which gives a good impression of the arrangement of the blood supply. Osteoblasts from the periosteum produce a layer of woven bone on which mesenchyme cells as well as a network of blood cells are positioned. Further woven bone encloses the vascular network with osteogenic cells which become osteoblasts and lay down lamellar bone so that the enclosed cavities become greatly reduced. This process is repeated, but there seems to be a hiatus in the production of the next successive deposition of woven bone. There is a so-called bright line, separating each lamina, which stands out because the canaliculi of the osteocytes do not pass across it (fig. 10.3a). This means that the osteocytes, which may be in close proximity in the woven bone, derive their nutrients from separate sets of blood networks. The distance from blood vessels of the furthest osteocyte is 0·13 mm; 90 per cent are 0·10 mm away.

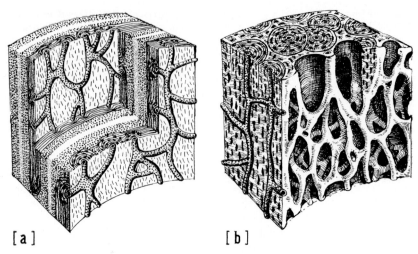

[a]   [b]

Fig. 10.3. Block diagrams of cortex of long bones. (*a*) Laminar bone (after Currey), (*b*) Haversian bone (after Ham).

Currey has calculated the volume of vascular canals as 1·40 per cent and the internal surface area of these channels as 32·9 cm^2 per cm^3. Currey has compared the structure of laminar bone to a multistorey hotel with many corridors on the separate floors and only a few lift shafts, an analogy that aptly describes the blood supply situation. From these studies, it is evident why this type of bone is by far the most common among terrestrial vertebrates. It is most efficient with regard to its blood supply, and furthermore its strength is extremely good. In spite of laminar bone being so widely distributed, it has received scant attention in textbooks. Emphasis has always been given instead to Haversian bone. Haversian systems in bone consist of longitudinal cylinders of bone with concentric lamellae (fig. 10.3*b*). In the central canal (the Haversian canal) there are blood vessels, and in vertical sections there are canals that link one Haversian system to another; these Volkmann's canals also contain blood vessels.

The periosteal surface is not smooth but is thrown into fine longitudinal corrugations, and these can be accentuated by the osteoblasts so that a longitudinal canal becomes enclosed. Osteoblasts line the surface and begin the production of successive lamellae of bone. As osteoblasts are incorporated as osteocytes, so further osteoblasts are recruited from the enclosed mesenchyme. Eventually the calibre of the tube is reduced until it houses only blood vessels. The Haversian system is delimited by a marked line, the cement line. The compact cortex is made up of bundles of these bony cylinders.

When transverse sections of Haversian bone are examined under the microscope, it becomes evident that some of the osteones cut across one another. This means that tunnels have been eroded out of the pre-existing bone so that new Haversian systems could be laid down. In fact, Haversian bone exhibits a high degree of bone turnover. Haversian systems are being produced all the time, even though this involves the continual destruction of older ones. Such a high degree of turnover allows the bone as a whole to have its shape modified to meet the changing requirements of the organism.

Animals that in the young stages do not possess Haversian bone tend to develop it as they get older. Currey has condemned the notion current in many textbooks that Haversian systems are the basic units of compact bone. To suggest this is tantamount to describing wrinkles as the basic units of skin!

When comparisons are made between laminar and Haversian bone it is evident that Haversian bone is less efficient. The interstitial bone, that is the remnants of previous Haversian systems that are left between the new, have their osteocytes virtually cut off from the source of nutrients, as they are on the wrong side of the cement lines. Even without the barrier of cement lines, these cells are up to 0·14 mm from the blood vessels; this is appreciably worse than in laminar bone. When it comes to the degree of vascularization, the comparisons are more striking. In Haversian bone the surface area of vascular channels per cm^3 is only 26·6 cm^2 compared to 32·9 cm^2 in laminar bone. It is quite evident that laminar bone is more efficient as well as stronger than Haversian, yet in the life of an individual Haversian replaces laminar.

Laminar bone for all its efficiency is not adaptable. There is hardly any way in which remodelling can occur. Any disturbance or disruption of the vascular supply results in the death of the adjacent osteocytes. The consequence of this is that osteoclasts will remove the offending area and this will, of necessity, involve boring out a channel. A Haversian system will thus be introduced into the laminar bone. The effect of this will be to further disrupt the vascular supply, there will be further cell death or necrosis, and osteoclasts will be induced to remove yet more material. In this way a chain reaction develops which ensures the eventual replacement of laminar bone by Haversian. All the stages in this process have been documented in one of the giant herbivorous dinosaurs (fig. 10.4).

With regard to the respective virtues of Haversian and laminar bone, in the final analysis a balance has to be struck between higher efficiency and adaptability, and it is the former that is generally sacrificed for the benefits that accrue from the latter.

10.4. *Acellular bone and cementum*

The bone of some of the most highly advanced living bony fish does not include bone cells. Here again, as with the heterostracan ostraco-derms, there is a bone-like tissue without osteocytes, and in teleost

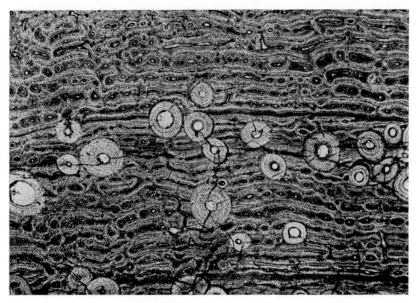

Fig. 10.4. Micrograph of transverse section of cortex of femur of Jurassic herbivorous dinosaur, showing typical laminar bone with Haversian systems penetrating and thus disrupting the original pattern, × 20 (photo J. R. Mercer).

fish, it is possible to trace its gradual evolution from a typical fish bone with included osteocytes. Fish osteocytes are much more variable than those found in the higher vertebrates, but the typical form is similar to the one illustrated in Chapter 7. From bone with such osteocytes, it is possible to trace a transition to a bone in which most of the cell bodies are confined to the vertical ascending passages termed the canals of Williamson with only the branching cell processes extending into the bone tissues. Thereafter, there is bone in which even cell processes do not penetrate. There is no doubt whatsoever that the acellular bone of modern teleost fish has evolved from the cellular variety. It is known that this acellular bone is less readily able to remodel than its cellular version. It is not understood why some of the most advanced of living aquatic vertebrates should have evolved acellular bone. On the face of things this should be disadvantageous. However, there it is.

In the mammals a further type of acellular bone is developed. This is generally only present as a covering around the roots of the teeth, although in all the major herbivorous mammals it is also present on the crowns. Although this tissue is clearly a type of bone, it is designated cementum. Embedded in this tissue are bundles of collagen fibres that are also embedded in the bone of the jaws. This arrange-

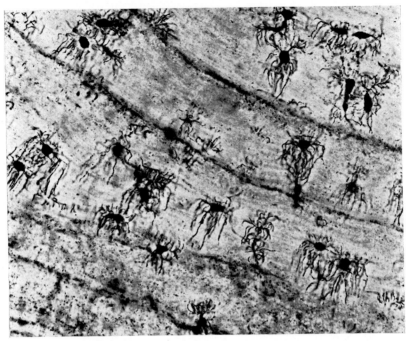

Fig. 10.5. Micrograph of human cementum with canaliculae drawn out towards surface of new apposition, × 400 (photo J. R. Mercer).

ment provides a specialized fibrous joint by means of which teeth are held in their socket but at the same time are allowed a certain degree of movement, albeit slight.

Cementum is divided into two types, that nearest the crown and that towards the apex of the root. The former is entirely acellular, the latter has included cells which are termed cementocytes; they can be readily distinguished by the fact that there are few if any cell processes enclosed in the tissue on the surface of apposition but the cell bodies become incorporated in the matrix they have secreted. As further matrix is produced the cell processes on the surface of new deposition are drawn out as if in an attempt to maintain their connection with the source of nutriment (fig. 10.5). Cementocytes are dispersed throughout the matrix, and are not normally in contact with one another via their canaliculi. Examples are on record where the cell body itself has managed to migrate to keep pace with the formation of new matrix so that it did not become enclosed for quite some time. Although cementum is clearly a highly specialized type of bone, the particular modifications that are observed are not at all well understood. It is noteworthy that in a group of newly discovered fossil chimaera fishes,

71

there are pharyngeal teeth around the roots of which there is a cementum-like tissue. This is a remarkable example of parallel evolution since this particular tissue cannot possibly be directly related to the cementum of mammals.

One of the issues which leads to a certain amount of confusion in the study of bone tissue throughout the vertebrates is the way in which similar structures evolved to subserve similar functions. For example the laminar bone of dinosaurs and that of cattle are indistinguishable, yet these two groups are in no way genetically related and have developed identical types of bone quite independently.

CHAPTER 11
dentine and enameloid

11.1. *Armour dentine*

In the early vertebrates, the outermost layer of the dermal armour is thrown into ridges or tubercles, and the pattern of this ornamentation is very characteristic for the different genera and species. When sections are made of this material it can be seen that the tubercles are penetrated by a system of tubules which appear to radiate out from a central cavity. These tubercles are fused to the underlying aspidin. In many instances the tubules divide up into fine terminal tufts at the periphery and there are also fine lateral branches (fig. 11.1).

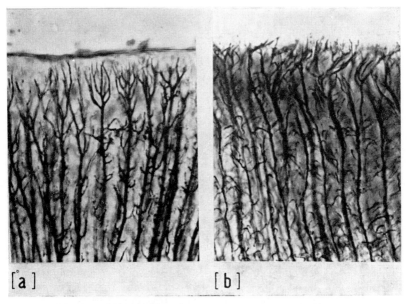

[a]   [b]

Fig. 11.1. Micrographs of dentine showing terminal tufts and lateral branches of dentine tubules, (*a*) heterostracan dermal armour, (*b*) human tooth × 500 (photos J. R. Mercer).

This structure of tubules penetrating the hard tissue is identical to the dentine of teeth. From this it is reasonable to deduce that it must have been formed in a comparable manner.

11.2. *Formation of dentine*

Dentine is a mesodermal tissue (strictly speaking it should be termed ectomesenchyme because the odontoblasts are derived from the neural crest). The outline of the tooth, tubercle or scale is defined by epidermal tissues beneath which mesenchyme cells aggregate. As the formation of dentine has been most intensively studied in man and rat this account is based on these two animals. Nevertheless, there is every reason to accept that this pattern applies to all dentine throughout the vertebrates.

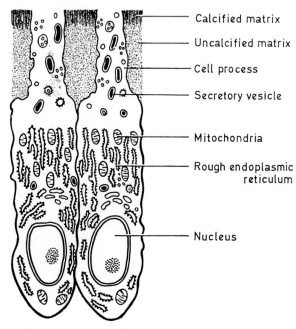

Calcified matrix

Uncalcified matrix

Cell process

Secretory vesicle

Mitochondria

Rough endoplasmic reticulum

Nucleus

Fig. 11.2.   Diagram of odontoblast (after Pindborg and Matthiessen).

To begin with the odontoblasts line up against the basement membrane formed by the epithelial cells. Cell processes extend towards the membrane, the nuclei being located at the opposite pole of the cells. There is a well developed rough endoplasmic reticulum which is indicative of the synthesis of protein for export (fig. 11.2). The cells of the subondontoblast layer stain for alkaline phosphatase and also have a rough endoplasmic reticulum. The subodontoblast cells are responsible for the secretion of the initial collagenous fibres of the organic matrix, known as von Korff's fibres. The initial matrix becomes highly mineralized and this first calcification seems to be due to the activity of the subodontoblast layer. Thereafter, this role is taken over by the odontoblasts themselves. The branching cell processes are surrounded by matrix, and as further matrix, mainly in the form of collagen fibrils, is secreted, the odontoblasts retreat leaving a wide tubule along which the major cell process extends (fig. 11.2). The collagen fibrils are spread on to the earlier matrix in consecutive layers all parallel to one another. In each layer the fibrils form a criss-crossing meshwork (fig. 11.3). As the odontoblast retreats the cell body itself always avoids being incorporated into the matrix. As the development of dentine proceeds, the odontoblasts come into closer contact with one another and become columnar in shape.

The calcification takes place in a calcifying front with spheres of radiating mineral or calcospherites being produced (fig. 11.4). Occasionally there are areas termed interglobular dentine where the dentine has

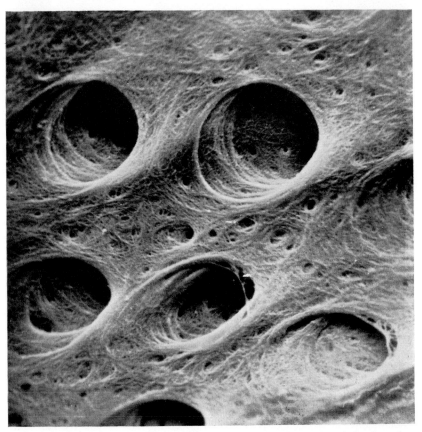

Fig. 11.3. Scanning electron micrograph of collagenous matrix of dentine, the interface between the odontoblasts and predentine exposed by removing the odontoblasts and pulp × 17 000 (photo supplied by Dr A. Boyde, University College, London).

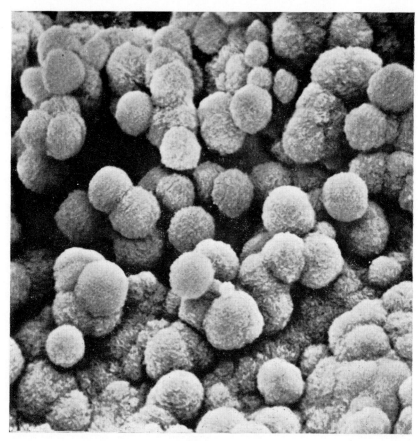

Fig. 11.4. Scanning electron micrograph of calcospherites in elephant molar dentine, with the mineralizing front exposed by dissolving the collagenous predentine (photo supplied by Dr A. Boyde, University College, London).

not been fully calcified. In such areas the spherulitic calcification is very prominent. The dentine tubules are continuous through these areas.

There has been a long controversy on the nature of the material within the dentine tubules. In the part of the tubules nearer the pulp there is a highly mineralized zone, the peritubular dentine, which has little collagen in it but is composed of a calcified proteoglycan matrix. It seems reasonable to suggest that the lining of the tubules contains chemically inert protein–polysaccharide complexes in exactly the same way as the lacunae and canaliculi of bone. In the dentine of the fossil heterostracans the tubules are also lined with a chemically inert polysaccharide.

11.3. *Original role of dentine*

This highly organized tissue formed the surface ornamentation of the dermal armour and scales of jawless vertebrates. These animals had not evolved teeth but were covered in dentine, which is the main component of vertebrate teeth. The epidermal tissue, which covered the tubercles in the early developmental stages, was sloughed off, apart from a surviving network around the bases of the tubercles. The dentine must have formed the main barrier between the animal and its water environment.

A vital role in any tissue acting in this capacity is sensitivity and it has been suggested that this was the original role of dentine. The organization of tubules permeating the hard tissue would thus be taken to have developed to act as a route of sensation. Although nerve fibres have been demonstrated passing up some dentine tubules from the pulp cavity, they are insufficient to account for the observed sensitivity experienced in, say, human dentine. Substances known to affect nervous tissue elicit no response when applied to dentine, whereas heat, cold, air and sweet solutions all have marked effects. It is now recognized that the sensitivity of dentine is occasioned by physical pressure down the tubule to the pulp. In an aquatic medium, a dentine-like tissue could be sufficient to pick up such stimuli. This plausible theory to account for the basic structure of dentine cannot be the whole story, because some of the tubercles have an outer cap through which the tubules do not run.

An animal's external covering must be capable of regeneration if the animal is not to be endangered by every small wound. For the heterostracans many details of the healing of the armour have been described. Where a fracture has taken place or even a piece of the armour removed accidentally or by reason of the attentions of a predator, the break is sealed over by a new set of dentine tubercles (fig. 11.5*a*). This means that cells from the surviving network of epidermal tissue must have spread over the lesion and that new odontoblasts began the process of dentine formation.

In areas of the carapace which suffered intermittent abrasion or undue irritation of some sort or other, new tubercles developed on top of the old. There are examples of worn tubercles that have new globular growths of dentine overlapping them. There are a number of specimens from Russia where the tubercular surface is entirely covered by a thin sheet of new dentine. To understand how the new dentine could have arisen in these positions it is necessary to postulate proliferation of epidermal tissue so that it overwhelmed the old tubercles. Within such blisters, odontoblasts were differentiated and new dentine tubercles formed. In some instances, as the new tubercles were growing, the underlying old tubercles were actively resorbed by osteoclasts or 'odontoclasts' (fig. 11.5*d*).

In the armour of all the early vertebrates the succession of tubercles

77

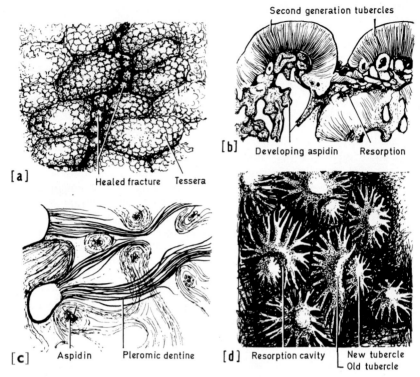

Fig. 11.5. Healing and regeneration in heterostracan armour dentine. (a) Ornamentation of superficial tesserae with fracture sealed by line of new dentine tubercles, (b) second generation tubercles with struts of developing aspidin and resorption of summits of first generation tubercles, (c) pleromic dentine filling spaces in spongy aspidin, (d) new dentine tubercles situated in resorption cavities on older tubercles (from L. B. Halstead, 1969, Calcified tissues in the earliest vertebrates, *Cal. Tiss. Res.,* **3,** 107–129).

always involved the new or second generation elements being positioned on top of the first generation elements (fig. 11.5b).

Since teeth evolved from dentine tubercles situated in the region of the mouth, it is reasonable to infer that the processes that occurred in dermal tubercles would be similarly present in teeth. It is common experience that each successive generation of teeth erupts from below the preceding one, the apparent reverse of the situation described in dermal armour. An examination of the embryological development of teeth reveals that this distinction is more apparent than real. Tooth germs are generated from epidermal structures—dental laminae—and as successive germs are budded off, they subsequently sink down by differential growth so that the new eventually come to be placed beneath the old. In fact the whole pattern of tooth replacement throughout the

78

vertebrates can be traced back to a simple healing mechanism in the skin of the earliest vertebrates.

In the later, more advanced, and considerably larger heterostracans that reached lengths of 2 m, the ventral parts of the carapace were subject to continual wear and the formation of blisters of soft tissue over worn parts would not have been easy to accomplish. In these instances the armour was strengthened by dentine filling in the vascular spaces in the spongy aspidin. This type of infilling dentine is termed pleromic.

The formation of pleromic dentine begins with the formation of odontoblasts where the epidermis covers the external openings in the spongy aspidin between tubercles. Differentiation of odontoblasts is presumably stimulated by abrasion at the summits of the tubercles. The first stage that is seen in the fossils is that these pores to the exterior are plugged by dentine. As wear proceeds, the odontoblasts retreat laying down dentine, so that the soft tissue contents within the aspidin are gradually replaced by dentine. The pleromic dentine has few dentine tubules but they are of considerable length and they wind through the anastomosing spaces of the aspidin (fig. 11.5c). The strengthened armour has a glassy appearance when viewed by the naked eye. It is difficult to see how the outer parts can receive any nutrients and so it is supposed that this tissue is gradually abandoned to its abrasive fate.

Normally in man, secondary dentine is formed in the pulp cavity of teeth as a compensation for the loss of tissue either by abrasion or disease. There is only a single recorded case of the cells of the pulp in armour dentine being reactivated to produce new dentine. This example was where odontoblasts entered the floor of the pulp cavity and the circumpulpar cells also began to produce new dentine. In the section which illustrates this point the normal dentine tubules suddenly turn through nearly 90° and run out of the pulp cavity into the aspidin. This one solitary observation may give some hints as to the very beginning of the development of secondary dentine in teeth.

11.4. *Mesodentine and semidentine*

Among two other groups of early vertebrate, the jawless cephalaspid ostracoderms and the jointed-necked, armoured arthrodires, primitive jawed vertebrates that gave rise to the sharks and their allies, there are two further types of dentine which have been used as evidence of the origin of dentine from a bone-like tissue. Among the primitive cephalaspids the dermal armour is thrown up into tubercles that are composed of bone, in which are included bone-cell spaces with interconnecting canaliculi. At the outer margin of the tubercles the canaliculi are aligned normal to the surface. When the geological history of the cephalaspids is traced, these outermost canaliculi become greatly elongated and give an appearance of dentine (fig 11.6 *a,b*). This type of tissue has been termed mesodentine by T. Ørvig. It has been contended that this tissue illustrates the way in which dentine evolved from bone, but it is more likely to be a parallel evolutionary development of a dentine-like tissue within a single group of animals. On the other hand the tissue of the tubercles of arthrodire armour named semidentine by Ørvig is undisputedly a form of dentine. In this the dentine tubules are typically developed with their fine lateral branches. They were formed by the gradual retreat of the odontoblasts as in all dentine, with the one major difference that the cell bodies were generally trapped in the matrix which

 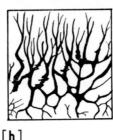 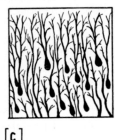

[a]　　　　　　　　[b]　　　　　　　　[c]

Fig. 11.6. Mesodentine and semidentine. (*a*) Primitive mesodentine with interconnecting canaliculae, (*b*) advanced mesodentine with dentine-like outer tubules, (*c*) semidentine with typical dentine tubules and with cell bodies enclosed in matrix (after Ørvig).

they secreted, thus becoming odontocytes (fig. 11.6*c*). The lacunae are ovoid and there are no canaliculi extending towards the pulpal surface. This type of dentine has been considered a half-way house between dentine proper, which does not have included cell bodies, and bone, which does. From the appearance of this tissue it is simply fairly typical dentine, except that the odontoblasts become trapped. It is difficult to attribute any special evolutionary significance to semidentine.

11.5. *Scale and tooth dentine*

There is a bewildering variety of dentine found in the teeth of vertebrates and the scales of fish, most examples being designated with their own special names. Dentine when it occurs in fish scales is known as cosmine and scales with an outer layer of dentine are called cosmine scales. In some of the teeth of advanced bony fishes the dentine does not possess normal tubules; instead the tissue is invaded by blood vessels and for this the name vasodentine has been coined (fig. 11.7*g*). In rays and skates, the dentine is organized into vertical units reminiscent of Haversian systems with a central pulp cavity. This type is sometimes termed osteodentine, and in some fishes the interstitial parts may contain osteocytes. This osteodentine (fig. 11.7*e*) has evolved in a single group of mammals, the tubulidentates, represented by the aardvark (*Orycteropus afer*). This tissue has arisen anew quite independently of that in the skates and rays.

80

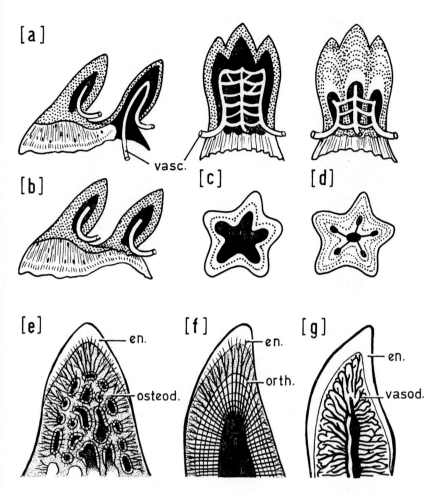

Fig. 11.7. (a–b) Adesmic cyclomorial scale in vertical section, (a) initial dentine unit showing addition of new unit at right, (b) further stage showing fusion of basal plates. (c–d) Polydesmic synchronomorial scale formed by fusion of six lepidomoria, in vertical (upper) and transverse (lower) section, (c) early development showing anastomosing blood vessels and thin outer shell of dentine, (d) further stage to show central pulp cavities of six lepidomoria, a central one with five surrounding. (e) Shark tooth in vertical section, to show outer enameloid cap and underlying osteodentine. (f) Shark tooth in vertical section to show normal or orthodentine. (g) Teleost tooth in vertical section to show outer enameloid cap and underlying vasodentine without tubules but invaded by a complex system of blood vessels. en, enameloid; orth, orthodentine; osteod, osteodentine; vasc, blood vessels; vasod, vasodentine (after Ørvig).

81

The relationship between normal dentine or orthodentine and osteodentine has been elucidated by Ørvig's study of the scales of primitive extinct sharks. The simplest elements or lepidomoria of the scales comprised a single vascular loop contained in a simple pulp cavity, the dentine crown was fused to a bony basal element. Thereafter, subsequent lepidomoria were added on, sometimes with crowns remaining separate (fig. 11.7a,b) but frequently in direct contact so that the outer wall of one would have been shared with the successive lepidomorium. Such areal growth is termed cyclomorial, and contrasts with the next evolutionary stage where the individual lepidomoria appear simultaneously, a type of growth known as synchronomorial. However, the individual lepidomoria can still be recognized as they retain their own pulp cavities (fig. 11.7c,d). It can be seen that apparently simple scales were evolved from others which appeared complex.

In the same way it is likely that osteodentine is a primitive tissue that became more simple during the course of evolution and gave rise to normal dentine or orthodentine (fig. 11.7e,f).

11.6. *Enameloid*

Covering the tubercles of the Ordovician heterostracan *Astraspis* there was a glassy cap which has been identified as enamel. This tissue in well preserved sections is seen to have tubules penetrating it and is now recognized to be a type of dentine sometimes called vitrodentine. When viewed under polarized light, this capping is in optical continuity with the underlying tissue of the tubercles. In later genera of heterostracan ostracoderms the fine terminal branches of the dentine tubules clearly penetrate the outermost layer of the tubercle. Yet when these sections are viewed in polarized light the outer margin is seen to be highly mineralized and forms a brilliant white cap. The surface appearance of the tubercles is shiny and it is evident that the initial formation of dentine differs from that which develops subsequently.

This outermost portion has been designated enameloid which simply means 'like enamel'.

The shiny covering of fish teeth is also generally labelled enamel, because it looks like it, but unlike true enamel it is mesodermal in

Fig. 11.8. Fish scales. (a) Primitive elasmobranch scale showing areal or cyclomorial growth of polydesmic units each with its own basal bony layer, (b) advanced elasmobranch scale or placoid scale with complex pulp cavity. (c) Primitive crossopterygian fish armour showing successive generations of tubercles of dentine and enameloid, (d) tubercles with enameloid capping lost, (e) advanced stage with dentine lost, (f) typical cosmoid scale with layer of dentine (cosmine) and thin enameloid capping. (g) Primitive actinopterygian, chondrostean or palaeoniscoid, scale with thick enameloid (ganoin) capping, middle dentine and basal bony layer, (h) holostean scale with reduction of dentine (cosmine) layer, (i) advanced actinopterygian, teleost, with only basal bony (isopedin) layer surviving, (j) teleost cycloid scale showing concentric growth rings, (k) teleost ctenoid scale with fine spicular ornamentation (after Ørvig).

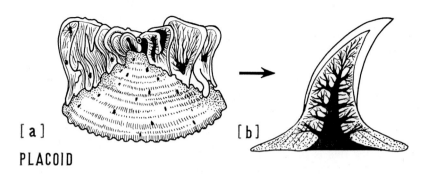

[a]
[b]
PLACOID

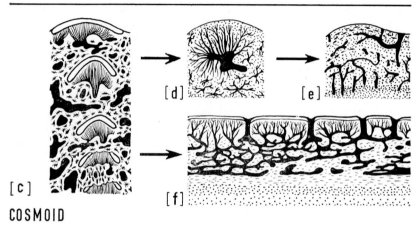

[c]
[d]
[e]
[f]
COSMOID

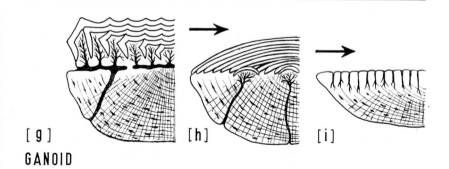

[g]
[h]
[i]
GANOID

[j]
CYCLOID
[k]
CTENOID

origin; again the term enameloid would seem to be the most appropriate. In the development of both the teeth and the scales of sharks the shape of the crown is determined by the epidermal cells. Within this defining cap, the dental papilla with its dentine-forming cells becomes organized. The first stage of tooth formation is that the future odontoblasts secrete the organic matrix in which there are but few collagen fibrils, and into which, moreover, cell processes do not extend. This matrix becomes highly calcified and forms a clear glassy cap to the tooth. Once this is produced dentine formation proceeds in the normal way. Both the method of formation and the type of cells responsible establish that this capping tissue is not true enamel, which is an ectodermal tissue. Fundamentally it is a kind of dentine serving the function of enamel.

Perhaps one of the most striking observations regarding the surface of vertebrate teeth was made by T. Kerr. He showed that in newts and salamanders, which possess teeth in the tadpole stage, the outer tissue of the teeth was mesodermal in origin and hence enameloid. On metamorphosis the covering was produced by ectodermal cells and was in fact true enamel. The reason for this dramatic change is something that still eludes workers in this field.

11.7. *Fish scales*

The placoid scale of the sharks is generally considered to be the most primitive and simplest type of scale, although early authors believed that it was derived from the gradual break-up of the dermal armour of the early vertebrates. It has now been established that the placoid scale is the end product of a long evolutionary history. The scales of the first sharks were highly complex, being formed by the successive cyclomorial addition of separate units each contributing to the bony base (fig. 11.8a). Eventually this type of scale gave way to the typical placoid scale (fig. 11.8b), the only evidence of its history being in the ridges of the crown and the complex branching of the pulp cavity.

The scales of primitive bony fishes are divided into two contrasting types. The cosmoid scale has a smooth surface with minute pores which give the scale a matt surface to the naked eye. There is a thin enameloid capping but the main tissue of the superficial part of the scale is dentine (termed cosmine in fish scales). The cosmoid scale is characteristic of the lungfish and crossopterygian fishes, but among the early representatives one find scales composed of dentine tubercles with thick enameloid caps (fig. 11.8c). It is possible to trace two separate evolutionary lines; one leads to the typical cosmoid scale (fig. 11.8f) whereas in the other the enameloid is gradually lost (fig. 11.8d), and then the dentine, so that eventually the scale is composed only of bone, termed isopedin among the fish (fig. 11.8e).

Ganoid scales are so called because of their thick enameloid covering, known as ganoin. The most primitive types are made up of the three tissues bone (isopedin), dentine (cosmine) and enameloid (ganoin) (fig. 11.8*g*). The basal bone is penetrated by Sharpey's fibres, bundles of collagen that anchor the scale to the underlying dermis. There are also ascending vascular canals of Williamson, which are sometimes known as lepidosteal tubules which become prominent in the later ganoid scales where the dentine is reduced and finally lost (fig. 11.8*h*). In the final stage in the evolution of the ganoid scale the outer enameloid layer is lost and only the bone remains (fig. 11.8*i*). In this last type of scale Sharpey's fibres also penetrate the scale from its upper surface, as the scale is entirely embedded in the skin. These advanced scales are of two kinds cycloid (fig. 11.8 *j*) or ctenoid (fig. 11.8 *k*); the names merely describe their gross appearance (Greek: kuklos, a circle; ktenos, a comb).

The overall pattern of change in fish scales is one of gradual reduction of the hard tissues, which produces a lightening of the armour. This is correlated, in the advanced bony fish, with the presence of hydrostatic organs, the air bladders, perhaps developed by gradual transformation of the accessory breathing organs or lungs of the primitive fishes.

CHAPTER 12

enamel

12.1. *Formation of enamel organ*

Tooth development in the mammals begins with the proliferation of the epidermal cells of the oral epithelium. This forms an intucking, the tooth bud, which is connected to the oral epithelium by a band of similar cells, the dental lamina. In amphibians and reptiles this structure is persistent, and successive tooth germs are budded off from it, but in the mammals only one tooth germ is produced from the dental lamina per tooth locus. The permanent teeth, however, are budded off from the germ of the corresponding milk or deciduous teeth, a situation quite distinct from that in reptiles.

The tooth bud develops a slight dent beneath which there is a concentration of mesenchyme, which is the dental papilla. This stage of development is the cap stage and it gives way to the bell stage (fig. 12.1).

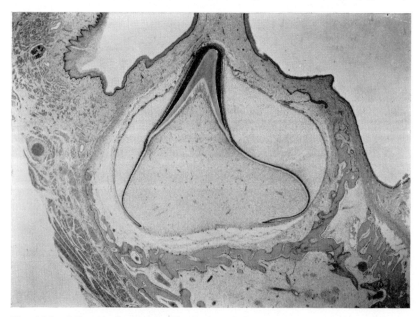

Fig. 12.1. Micrograph of enamel organ at bell stage, showing inner and outer dental epithelium, stellate reticulum, mesodermal papilla, ameloblasts, enamel, dentine, predentine and odontoblasts (photo J. R. Mercer).

The outer surface of the enamel organ is made up of cells termed the outer dental epithelium; those enclosing the dental papilla form the internal dental epithelium, which determines the shape of the tooth. The greater part of the enamel organ is formed by cells that form the stellate reticulum. The odontoblasts, lined up against the internal dental epithelium, then begin the production of the organic matrix of dentine. Once dentine formation has been initiated, enamel begins to be laid down. Once odontoblasts are differentiated the cells of the internal dental epithelium become elongated into long columnar cells with a prominent rough endoplasmic reticulum and numerous ribosomes at the secretory pole, and the nucleus and mitochondria at the end adjacent to the cells of the stratum intermedium of the stellate reticulum. Unlike the protein synthesizing and secreting cells already discussed, the ameloblasts or enamel forming cells do not operate in an environment of vascularity. However, the pre-ameloblasts are rich in glycogen, and this must provide the source of energy for enamel formation or amelogenesis.

12.2. *Amelogenesis*

The enamel forming cells, the ameloblasts, are ectodermal in origin. The cells have a prominent rough endoplasmic reticulum characteristic of protein synthesizing and secreting cells (fig. 12.2). At the margin where amelogenesis begins, secretory vesicles are observed and this zone develops into a large process, the Tomes process, which is differentiated from the rest of the cells by a curtain of thin tonofilaments. The ameloblasts secrete the enamel matrix, the nature of which as noted in Chapter 2 is not at all well understood. Alkaline phosphatase is present in the cells of the stratum intermedium.

Developing enamel has a honeycomb appearance, with the Tomes processes fitting into the 'cells'. There are also electron-dense granules in the secretory vesicles.

The crystallites of apatite are deposited in the organic matrix by the Tomes processes, appearing first as fine filaments and gradually achieving their full girth. The apatite crystallites of enamel are enormous in comparison with those found in bone and dentine, having a length of 160 nm, width of 40 nm and thickness of 25 nm. This contrasts with bone crystals 45 nm in length and width with a thickness of only 2·5 to 5 nm.

With maturation of enamel the crystallites grow and the organic matrix is squeezed out. As already noted there is a significant change both in the amount of organic matrix and in the relative amino acid composition of the remaining organic residue. At the same time the ameloblasts, so far studied in rats, undergo further changes which can be termed the resorptive phase. Numerous vesicles occur which seem to be concerned with the uptake of material, numerous mitochondria are present in the same area and there is a marked development of

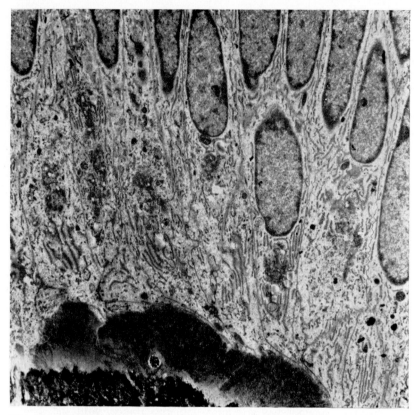

Fig. 12.2 Electron micrograph of ameloblasts showing Tomes processes and developing enamel prisms at the amelodentinal junction × 3500 (A. F. Hayward).

lysosomes, which are responsible for the digestion of ingested materials.

Under the ordinary light microscope mammalian enamel is seen to be organized into a series of prisms. In cross section under the electron microscope these prisms look like linguoid ripples, or horse's hooves bounded by a thin organic substance, the prism sheath (fig. 12.3).

The Tomes processes are conical in shape and crystallites are shed from them oriented at right angles to the plasma membrane. At the apex of the Tomes process a minute account of organic matrix is enclosed in the enamel and this forms the prism sheath. At the apex of the Tomes process the crystallites are laid down at almost 90° to one another so that the light refraction accentuates these areas which are the boundaries of the prisms. Because the matrix and mineralization produce an irregular surface the retreat of the ameloblasts tends to be at an angle. It is as if the ameloblast slides up one side of the newly formed enamel. It is this relative movement of the ameloblasts that gives the enamel prisms their characteristic appearance in section.

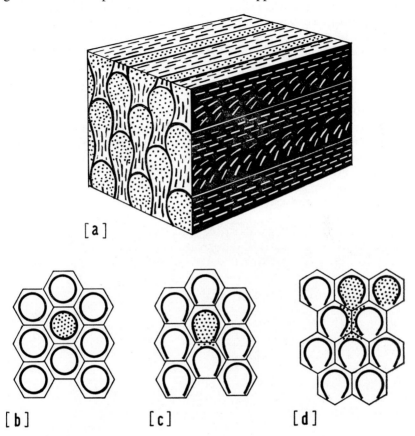

Fig. 12.3. (*a*) Block diagram of the arrangements of enamel crystallites in prisms from human enamel. The different appearances seen under the light microscope can be attributed to optical effects consequent upon this packing. (*b–d*) Diagrams of cross-sections of enamel prisms (hexagons) to illustrate contributions of a single ameloblast (stippled), heavy circles or horseshoes represent organic inclusions, (*b*) single ameloblast contributes the crystallites for a single prism, the situation in bats, insectivores, whales and lemurs, (*c*) crystallites contributed by two ameloblasts, i.e. each ameloblast produces crystallites which become part of two separate prisms, as in rabbits, hares, marsupials and ungulates, (*d*) each prism receives contributions from four ameloblasts, as in man, carnivores and elephants (after Boyde).

The pattern of the arrangement of the Tomes processes relative to the arrangement of enamel prisms differs from one group of mammals to another. There are three basic arrangements which are illustrated in fig. 12.3. The heavy lines mark the major changes in crystal orientation, which are the areas of the prism sheaths. In the first type, characteristic of bats, insectivores, toothed whales and lemurs, a single ameloblast is responsible for a single enamel prism (fig. 12.3b). In the second, an enamel prism receives contributions from two ameloblasts (fig. 12.3c). This type of enamel is found in rabbits, hares, marsupials and ungulates. The third type of enamel prism receives contributions from four cells (fig. 12.3d). Elephants, carnivores and man possess the last kind of enamel.

12.3. *Structure of enamel*

Mammalian enamel is composed of enamel prisms, which are easy to observe under the microscope. When viewed in electron micrographs the prisms are seen to be made up of enormous numbers of closely packed crystallites (fig. 12.4).

Although the general direction of the prisms is from the dentine surface to the outer surface of the tooth, in detail the prisms take up

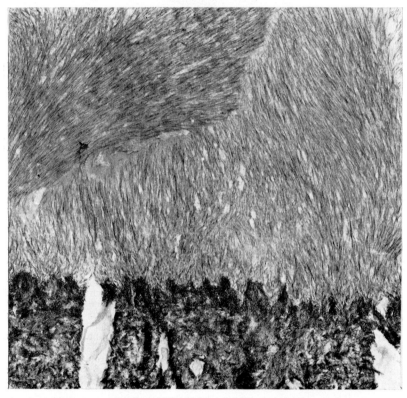

Fig. 12.4.   Electron micrograph of apatite crystallites in adjacent enamel prisms at the amelodentinal junction, × 15 000 (A. F. Hayward).

a wavy course and in certain areas become extremely twisted to give what is termed gnarled enamel. In the early stages of enamel formation, there appear to be odontoblast processes passing from the dentine into the first-formed enamel. These structures are known as enamel spindles. Further structures near the amelodentinal junction are enamel tufts that

Fig. 12.5. Micrograph of enamel to show arrangement of enamel prisms and enamel tufts at the irregular amelodentinal junction (photo J. R. Mercer).

are so called because they look like tufts of grass. These are organic leaves or sheets which seem to be thickened prism sheaths (fig. 12.5). They are less calcified than other parts of the enamel.

Under the light microscope two further sets of structure can be observed. The prisms, as they are traced from the amelodentinal junction to the surface of the tooth, turn either left or right. These directions are constant for alternate zones. This difference in orientation means that if longitudinal sections are viewed by reflected light one set will appear light and the other dark (if the direction of the light is reversed then the previously dark ones become light and vice versa). This optical effect of black and white banding is known as Hunter–Schreger bands. Finally there are the Brown Striae of Retzius which are incremental or growth lines. Each one indicates a slight hiatus in enamel formation.

12.4. *Evolution of prismatic enamel*

True ectodermal enamel is only present in land vertebrates. The prismatic type of enamel described is only known among mammals and then only the therian mammals, i.e. marsupials and placentals. No other form shows any sign of possessing enamel prisms. M. L. Moss pub-

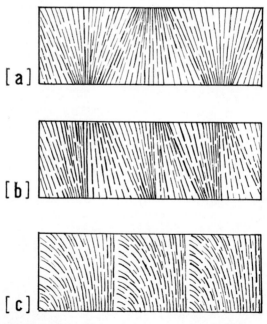

Fig. 12.6. Evolution of prismatic enamel. (*a*) Arrangement of crystallites in a sinusoidal pattern as in paramammals or mammal-like reptiles, (*b*) saw-tooth pattern in primitive Triassic and Jurassic mammals, (*c*) prismatic enamel of marsupial and placental mammals (after Moss).

92

lished a survey of fossil enamel, and was able to demonstrate that in the Permian and Triassic paramammals, or mammal-like reptiles, the orientation of the crystallites formed a sinusoidal pattern. It was suggested that in this situation the ameloblasts formed an undulating surface, with only poorly developed Tomes processes. The protoprisms, as with prismatic enamel, were produced by adjacent ameloblasts, but the retreat from the forming enamel surface was more exactly normal, i.e. at right angles (fig. 12.6a).

Among the earliest true mammals from the late Triassic period and the succeeding Jurassic periods, the orientation of the crystallites developed a degree of asymmetry, which can be attributed to a change in the behaviour of the ameloblasts. This new arrangement suggests that the ameloblasts were beginning to retreat at an angle thus foreshadowing the situation in prismatic enamel. This type of enamel has a 'saw-tooth' appearance (fig. 12.6b).

At the beginning of the Cretaceous period teeth of the first therian, *Pappotherium,* are known, and these have an inner zone of prismatic enamel. All subsequent marsupial and placental mammalian teeth possess the prismatic form of enamel (fig. 12.6c).

The asymmetry of the movement of the ameloblast and the resulting orientation of the crystallites reached a particular level at which it produced the optical effect which enables the prisms to be seen. At this point prismatic enamel appeared. The sudden advent of prismatic enamel is purely an optical effect; its evolution was in fact a gradual process.

CHAPTER 13

keratinous hard tissues

13.1. *Formation of keratins*

Keratinous tissues differ fundamentally from all those so far discussed. The epidermal cells responsible for the formation of keratinous tissues are protein synthesizing and retaining cells. In all cases the keratin is laid down within the cell. The cells become so loaded that their demise is assured—in a way they can be thought of as suicidal! There are two types of keratinous tissues; the soft keratins such as in mammalian skin, and the hard keratins of hair, nails, feathers, horn and baleen.

In the soft keratins the cells accumulate keratohyalin granules, whereas with the hard keratins microfibrils accumulate until the entire cell is transformed. In the soft keratins the keratinized cells desquamate, that is are sloughed off, and are continually replenished from beneath. Hard keratins contrast in that the structure produced, although dead, is continually growing from a generative zone. The structure of hair, horn, feather or baleen is relatively permanent. Apart from pathological conditions, it is only the hard keratins that become calcified.

13.2. *Hair and nails*

Perhaps the most intensively studied of all keratinous tissues is hair. An outline of its formation will serve as a guide to the basic processes involved, which will be applicable to all other keratinous hard tissues discussed.

As with the development of the enamel organ, the first stage begins with the proliferation of epidermal cells and their invasion of the dermis. At the deepest part of this penetration the epidermal tissue forms a cap enclosing a papilla of mesenchyme tissue. So far this is comparable to the early stages of the development of the enamel organ but here the similarity ends. The epidermal tissue differentiates into various parts; one area forms the sebaceous glands, but of more interest in this account is the formation of the hair itself. The main lining of the hair follicle is the external root sheath; at the base of the follicle the cells differentiate into the internal root sheath and the hair proper. The internal root sheath forms keratin, but of the soft variety, the cells accumulating keratohyalin granules. In the hair itself three zones can be recognized. There is a central medulla which is again a soft keratin, a thick cortex and a thin cuticle; both the latter are hard keratins (fig. 13.1). The base

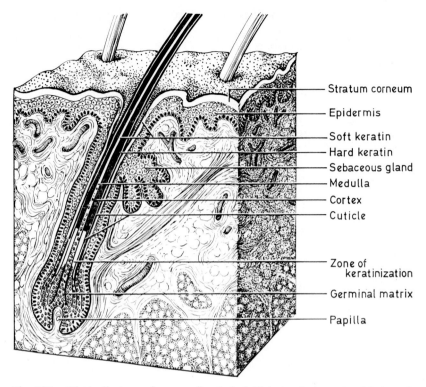

Stratum corneum
Epidermis
Soft keratin
Hard keratin
Sebaceous gland
Medulla
Cortex
Cuticle
Zone of keratinization
Germinal matrix
Papilla

Fig. 13.1. Block diagram of mammalian hair follicle to show parts of hair and types of keratinization (after Ham).

of the hair follicle forms the germinal matrix, and immediately distal to this is the zone of keratinization. As the cells produced by the proliferation of the germinal matrix are pushed further away from the papilla from which all the nutrients come, so they must die. During this process they synthesize keratin fibrils, and gradually all the cell contents become transformed into keratin. As the process continues the growing hair protrudes beyond the skin and becomes externally visible. Eventually the hair reaches its definitive length. At this point the base becomes club-like, and develops rootlets to attach it firmly to the follicle. The follicle itself shrinks and goes through a resting period. Subsequently the follicle develops the germ of a new hair which as it grows loosens the old hair which is then shed. In man this process is a continual one, but in most furry mammals of temperate regions such moulting is seasonal, a thick winter coat being exchanged in the spring for a lighter summer version, and the reverse in the autumn.

Nails and claws, which are more obviously hard tissues, also begin development from the ingrowth of epidermal cells into the dermis. The cells of the lower part proliferate to make up the nail matrix, whose cells divide and become keratinized to form the nail. With further cell division and nail formation the nail is pushed forward over a layer of epidermal cells, the nail bed. This process can continue until the distal part of the nail becomes free from the tissues of the digits. As with the hair the hard keratins are fibrous, the fibrils being aligned along the major axis of either hair or nail. In the claws of cats and rabbits there are marked zones of calcification, which maintain the sharpness of the claws because of the differential wear on the adjacent uncalcified parts (fig. 13.2).

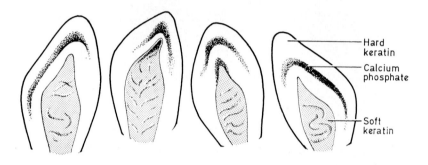

Hard keratin

Calcium phosphate

Soft keratin

Fig. 13.2. Diagram of rabbit claws to show distribution of calcium phosphate (after Pautard *et al.*).

13.3. *Feathers and scales*

Reptiles possess a covering of horny scales which are, like nails and claws, generated from the Malpighian layer of the epidermis. These are periodically shed as isolated flakes, but in the snakes the entire horny covering is sloughed off in one piece. Beneath the dead cuticle there are numerous chromatophores. These are responsible for the varied colour patterns among the lizards and snakes. The protective coloration of the chameleons is perhaps the most dramatic example of colour changes. The rainbow lizard, which is a strongly territorial animal, produces a variety of colour patterns but these are primarily for display.

Small plates of bone, osteoderms, are present beneath the dermis. These are best known in crocodiles. The surface-sculpturing of the bony scutes is reflected in the horny covering to give the characteristic pattern of crocodile skin.

Perhaps one of the most elaborate of dead keratinous structures is the bird's feather. In many respects the development of the feather is comparable to that of hair. Feathers all grow from their own follicles with the dermal papilla at the base. The structure of a feather comprises a hard hollow calamus or quill, at one time used for writing and today sometimes used as tooth picks. The quill has a hole at the base where the mesodermal papilla is situated, the main shaft or rhachis of the feather bears the vane, which comprises barbs or radii which are held together by barbules with their hooks or hamuli. The hooks on the anterior side of the barbules interlock with ridges on the posterior side of adjacent barbules (fig. 13.3).

If the radii become separated, they can be readily rejoined by preening. In the flight feathers there are friction areas where extra hooks are present to prevent the feathers from sliding too far apart.

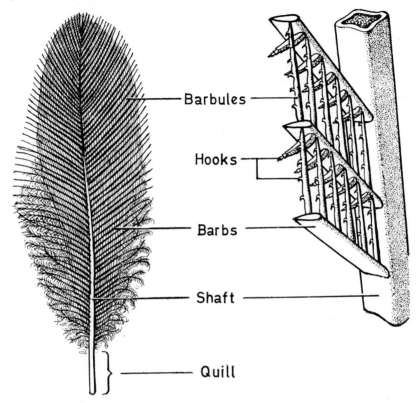

Barbules

Hooks

Barbs

Shaft

Quill

Fig. 13.3. Feather with detail to show interlocking of barbules by means of hooks or hamulae.

There are three main types of feather; the contour feathers or pinnae, which have all the structures already described, the insulating down or plumules in which the barbs remain free, and the hair-like filoplumes.

In all cases the feathers, like hair, are moulted, and new feathers are developed from the same follicles. The newly hatched bird is covered entirely with down, contour feathers developing only later.

The most noticeable feature of feathers is the wide variety of pattern and colour which they exhibit. The dun-coloured or mottled females frequently contrast with the dazzling colours of the male. Cryptic coloration typifies the females, while the males tend to be unduly conspicuous. The colours in feathers are produced by pigments within their substance, and also by reflection and diffraction effects. White is due to reflection, blue to the Tyndall scattering of incident light, while iridescent colours are the result of interference of light in the surface films.

13.4. *Antlers and horns*

Many herbivorous mammals sport striking outgrowths on their heads. These are either antlers or horns, and the two structures are entirely different in their formation; antlers are bony whereas horns are keratinous. The horn of the rhinoceros is composed entirely of numerous tubular hair-like filaments that are cemented together and attached to the nasal bones. In contrast, the horns of cattle, sheep and antelopes are made of keratin in the same way as nails and claws, and form hollow cones of varying shapes. The horny sheaths fit over bony cones which are outgrowths from the frontal bones—the *os cornu*. As with nails and claws, the horn is dead, but is produced from a germinal epidermal layer covering the bony horn core (fig. 13.4a). In this way the horny sheath grows throughout the life of the individual and becomes gradually thicker.

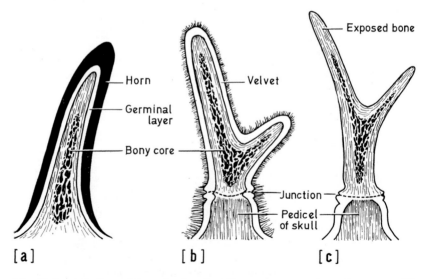

Fig. 13.4. (*a*) Section of horn to show keratinous cover over bony horn core, (*b*) developing antler covered in epidermal 'velvet', (*c*) bony antler after velvet has been shed (after Modell).

The North American pronghorn antelope is unique in possessing, as the name implies, a pronged or branched horn over a bony core. This animal is further remarkable in that the branched horny sheath is shed annually, a new sheath being formed beneath the old. The replacement horn begins its development from beneath the tip of the old and its complete growth and hardening do not take place until after the old one has been shed.

Perhaps the most primitive type of horn, from which it is believed both horns and antlers may have evolved, is that found in the giraffe. It is unbranched and is covered with hairy skin. The bony core is a separate ossification, which only later fuses to the skull. In some respects this type of horn is similar to the early stages of the development of antlers, where the bony growth is covered with a hairy skin.

True antlers are made of bone and are shed annually. In no sense are they related to any of the keratinous hard tissues, but they are dealt with here because they are sometimes confused with horns. At first the growing antler is covered with a hairy skin termed 'velvet', which has a rich blood and nerve supply and which is both tender and fragile (fig. 13.4b). Within this skin the formation of the antler takes place at an astonishing pace, comparable only to that of a bone sarcoma or primary bone cancer. When growth is complete, the antler ossifies, and the velvet dies and gradually peels or is rubbed off, beginning at the tips of the branches (fig. 13.4c). Once the velvet has been lost, the bearer becomes sexually aggressive. Towards the end of the breeding or rutting season, osteoclasts resorb the base of the antler so that it is no longer firmly attached to the pedicel of the frontal bone. In this condition the antlers are easily shed, generally accidentally, which occasions considerable surprise to the animal concerned and renders him of a much more humble disposition. There is little bleeding at this trauma and velvet grows over the lesion. The antler then begins its annual growth again.

Apart from the reindeer and caribou, only male deer produce antlers. In the first year the male fawn develops the pedicel on the frontal bone, the second year sees a simple spike. Thereafter each successive year witnesses the addition of further branches or points or tines. The lowest anterior point in red deer (*Cervus elaphus*) is termed the brow tine and subsequent points are known as bez, trez, royal and surroyal. The main trunk of the antler is termed the beam.

As with other skull roofing bones, the formation of antlers is by direct ossification; there is no cartilaginous template. Furthermore the cancellous bone does not contain any haemopoietic tissue.

The development of the heavy and often cumbersome antlers that are shed annually is one of the vagaries of evolution. These deciduous organs are of little use as weapons either of offence or of defence. Their main function is one of display, to intimidate rival stags and to impress the hinds with the owner's virility. When well matched stags are locked

99

in combat both adversaries frequently perish if they suffer the misfortune of getting their antlers entangled. However, the success in obtaining a harem may be a measure of the magnificence of the antlers and large antlers are presumably a result of sexual selection over many generations. It is known that stags without antlers, hummels, can also succeed in gaining harems against their antlered rivals. Hence this question is somewhat more complex than was previously thought.

In the ungulates with true unbranched horns, there is also a certain premium on sexual display, but at the same time the horns are frequently used as weapons for goring enemies. Rivalry between males involves the crashing of skulls or the bases of the horns against one another but rarely attempts at goring. Such behaviour would eliminate the most aggressive males and would inevitably lead to the demise of the species.

13.5. Baleen

The most highly calcified of all the keratinous tissues is baleen, which is popularly known as 'whale-bone'. Until the advent of plastics, baleen was a major support and contour moulder of the female of *Homo sapiens,* for many years being used commercially for the stays of corsets.

Baleen hangs in enormous sheets from the roof of the mouth of the 'whale-bone' whales, which feed by filtering small crustaceans, the shrimp-like krill, from the waters of the oceans. These sheets of baleen are fringed at their inner distal margins (fig. 13.5a). The plates of baleen are only a few millimetres in thickness and are about a centimetre apart. As with hair and nails they grow continuously being generated from epidermal tissue. Dermal papillae penetrate the plate and outfingerings of mesenchyme get incorporated into the material of the baleen plate. The plate is separated from the dermis by a thick layer of gum, the superficial surface of which is being continually sloughed off as the plate grows. The external surface of the baleen plate comprises a layer of keratinized covering horn. Sandwiched between the layer of covering horn are numerous horn tubes, which are themselves cemented together by a cementing horn (fig. 13.5b).

On the median distal edge of the plate—the lingual surface—the surface gets worn by the action of the tongue so that the covering and cementing horn is eroded away to expose the horn tubes. These may be as thin as threads of silk or like very stiff bristles. As well as trapping the krill, these fringe fibres of the horn tubes allow the food to be transferred to the tongue with ease. The horn tubes have a dermal papilla in their central cavity which, as the horn tube grows, disintegrates so that the canal merely contains a few bits of cell debris. The mesenchyme of the dermal papilla would seem to be the source of the calcification that takes place in baleen.

The horn tubes can be seen to comprise two zones, an inner tube bordering the horn cavity and an outer. The calcification is concentrated in the peritubular portion. The calcification is intracellular and is in alternating rings (fig. 13.5c). In electron micrographs, the keratinized epidermal cells can be seen to be divided into those in which calcification has proceeded far, and others which seem to be quite uncalcified (fig.

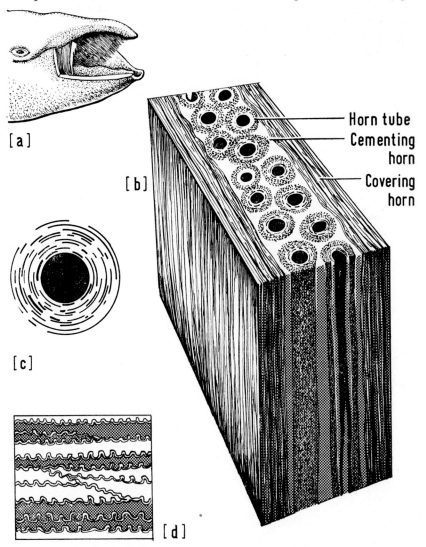

Fig. 13.5. Baleen. (a) Head of whale-bone whale with plates of baleen hanging from the roof of the mouth, (b) block diagram of baleen, (c) section of horn tube with areas of calcification indicated, (d) drawing based on electron micrograph of keratinized epidermal cells, those shaded being mineralized.

13.5*d*). From a purely functional point of view this arrangement would clearly give greater flexibility and strength to the horn tubes.

One aspect of the calcification of baleen that has been established is that no collagen is present; furthermore the crystallites of apatite seem to be comparable in size to those found in bone, dentine and cartilage. The intracellular calcification of keratinized cells is found in all terrestrial vertebrates, but only in baleen has it become a major aspect of the tissue to ensure its efficient functioning.

CHAPTER 14
eggshells and calcium carbonate deposits

14.1. *Eggshell formation*

A vertebrate hard tissue composed of calcium carbonate that is particularly well known in birds is the eggshell. This acts as a protective
covering for the developing embryo as well as an important source of
calcium for the skeleton of the chick. As the ovum moves from the
ovary down the oviduct, albumen is added and then the shell membranes.
When the egg reaches the shell gland, the uterus, calcification begins.
The first stage of the formation of the eggshell is the production of small
spherulites. These are composed of radial spicules of aragonite which
form around an organic nucleus. This organic material seems to be a
proteoglycan, a form of chondroitin sulphate. From these initial
spherulites, a zone of tabular or rhombohedral crystals of calcite grows
outwards from the organic core, but is prevented from expanding in
towards the egg because of the presence of the shell membranes. The
initial spherulites are called the mammillary knobs, and the further
calcification produces the cone layer of the shell. Thereafter, the calcite
crystals fuse to form the palisade layer which grows in an outward
direction; it is characterized by a herring-bone pattern caused by the
development of crystals diagonally across the incremental growth lines
(fig. 14.1). When viewed under polarized light, the different palisades
can be seen to be optically distinct. There seems to be some difference
between the organic matter of the palisade layer and that of the cone
layer. The columns of calcite of the palisade layer when viewed in
horizontal section interlock in a jigsaw-like manner. Scattered over the
shell are pores which run from the shell membranes between the palisade
columns to the exterior. These enable the developing embryo to take in
oxygen and give out carbon dioxide.

Once the shell is laid down a thin cuticle is deposited over the entire
surface.

14.2. *Evolution of eggshells*

The structure of the eggshell appears more variable when other groups
of egg-laying vertebrates are examined. The most primitive type of eggshell is found in turtles, tortoises, lizards and snakes. In these reptiles,
the basic unit of the shell consists of the mamillary knobs with their
radial aggregates of spicular crystals of aragonite (fig. 14.1). It is believed

that this arrangement represents the starting point for the later evolution of eggshells. The crocodiles, however, seem to have dispensed with this initial stage and their shells are made up entirely of the tabular aggregates of calcite crystals of the cone layer (fig. 14.1). Birds and dinosaurs possess all three elements, the mamillary, cone and palisade layers. In some dinosaurs as well as the ratites or flightless birds, the herring-bone pattern of the palisade layer may be poorly developed or even absent (fig. 14.1). The reasons for these differences are not understood.

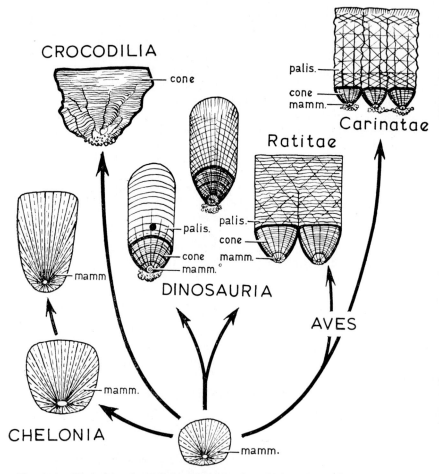

Fig. 14.1. Evolution of eggshells. Diagram of vertical sections of the basic units of eggshells to illustrate compositional differences of the various types of eggshell. Cone, cone layer of tabular or rhombohedral crystals; mamm., mamillary knob of radial spicules; palis, palisade layer of large crystals with herring-bone pattern (after Erben).

A tentative scheme of the likely evolution of eggshells is given in fig. 14.1.

In the study of dinosaur eggs, it has been found that the last of the dinosaurs, in the south of France, laid eggs in which there was a complete repetition of the shell wall. Such a pathological condition is known to poultry farmers and is due to the deficiency of the hormone vasotocin. This deficiency results in the egg moving to and fro between the shell gland and the oviduct and thus receiving extra layers of shell. The embryo is effectively sealed off since the pores would be blocked and even if it were to develop it would be unlikely to be able to break through such a reinforced shell. It has, in fact, been suggested that this may be a contributory cause of the extinction of the dinosaurs.

14.3. *Supply of calcium for the eggshell*

The main source of calcium for the manufacture of eggshell comes from the skeleton. Among the reptiles, such as the marine turtles, the laying females show an appreciable reduction in their bones which become rarified during this period. This is likely to have been the case with the female dinosaurs but so far it has not been recognized. With birds and especially the laying hen the problems are somewhat different. In the hen the yolk contains a high proportion of calcium, yet this is being formed at the same time as the preceding egg is receiving its shell. The bird has to be able to mobilize its calcium extremely rapidly. In fact 10 per cent of a bird's skeleton can be mobilized in 24 hours. For this birds have evolved a specialized type of secondary bone known as medullary. This bone is laid down as fine spicules which ramify through the marrow of the long bones prior to the formation of the eggshells. The dense population of osteoblasts is rapidly converted to one of osteoclasts and the medullary bone is quickly removed. The calcium level in the blood is greatly increased in consequence and ends up in the vascular supply of the egg gland.

14.4. *Supply of calcium for the embryo*

While the embryo is developing during the incubation of the egg, the skeleton is formed. This requires large amounts of calcium. In 1822 William Prout published analyses of the content of hens' eggs and recorded the astonishing fact that the amount of calcium increased fourfold. Even the yolk from which the embryo would have acquired the calcium seemed to increase the amount it contained. Prout could not see how the increased calcium could have been derived from the shell as there were no blood vessels in the vicinity and thought that, as there was plenty in the yolk, there would have been little point in using that from the shell. The problem was compounded by the fact that although the material for the skeleton was derived from the yolk, the amount in the yolk did not seem to decrease. As the egg and its shell are all that

105

there is, the ultimate source of the increase in calcium must be the shell. Without invoking a deity, this is the only possible explanation of these results. In the case of the hen this involves about 100 mg, or 5 per cent of the shell.

In the first few days of incubation there is a lowering of calcium in the yolk, but this then rises as calcium ions are removed from the shell. Just before hatching the yolk seems to be the main source of calcium. The chorioallantoic membrane, which comes to lie beneath the shell membrane, is concerned primarily with respiration. It has been suggested that the respiratory carbon dioxide leads to the solution of the calcium carbonate of the shell, as calcium bicarbonate, which is absorbed by the blood in the chorioallantoic membrane. Some support for this mechanism is provided by the fact that there is no bicarbonate in the freshly laid egg but the amount increases as incubation proceeds.

The region of shell resorption is at the base of the calcite palisade layers. The mamillary knobs and cones remain intact and firmly attached to the shell membranes, which later become detached from the shell. In a fresh egg it is difficult to detach the membrane from the shell.

Birds have solved the problem of providing the basic skeletal elements for their offspring in a novel manner. A special type of bone which can be readily mobilized is developed from which the calcium ions can be made available for the production of a shell of calcium carbonate to protect the developing embryo. As the embryo grows it is able to obtain the necessary calcium for its skeleton from that provided by the mother in the form of the eggshell. The supply of calcium for the skeleton of the developing young is accomplished differently in other groups of vertebrates.

14.5. *Endolymphatic sacs and calcium stores*

In nearly all vertebrates, there are deposits of calcium carbonate in the inner ear; in the lamprey and other agnathans these deposits are of calcium phosphate. The membranous labyrinth of the ear consists, in addition to the semicircular canals, of two main parts, a lower horizontal sacculus and a vertical utriculus (fig. 14.2), which contain areas of sensory epithelium; the saccular macula is oriented in a vertical plane and the utricular macula in the horizontal. Each sac contains calcareous deposits or statoliths, often generally described as otoliths, and these are in contact with sensory cells of the maculae. The otoliths, which are enclosed in the endolymph-filled sacs of the inner ear, are concerned with the sense of balance.

Extending from the junction of the sacculus and utriculus is a narrow duct which ends blindly in the endolymphatic sac, which is situated in the cranial cavity in contrast to the rest of the ear apparatus (fig. 14.2). In the primitive jawless vertebrates, the endolymphatic sac enters the cranium, and it contains particles of poorly crystalline calcium phos-

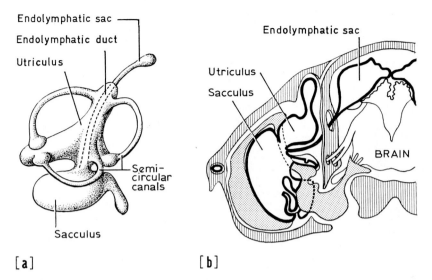

Fig. 14.2. (*a*) Utriculus and sacculus of amphibian middle ear, (*b*) section of amphibian head to show relationship of endolymphatic sac to other regions (after Simkiss).

phate. In the sharks and the cephalaspid ostracoderms the duct opens to the exterior. In the sharks and their allies the sac may be expanded just beneath the skin where it contains deposits of calcium carbonate.

However, it is in the amphibia that the endolymphatic sac reaches its most extreme development. In some genera it extends the entire length of the body. In the tadpole, the two sacs meet and fuse in the midline and grow posteriorly above the spinal cord. The endolymphatic sac develops outpocketings that protrude from the vertebral canal between vertebrae to form the calcareous sacs that are seen covering the spinal ganglia when the adult frog is dissected. Generally the amount of calcareous deposits in the tadpoles exceeds those of the adult and they may not in fact be present in all forms.

Normally during the life of the tadpole the skeleton is not ossified. There are sufficient calcium ions in the water and there seems every reason for the skeleton to ossify, yet it does not. Instead, large accumulations of calcium carbonate are formed in the endolymphatic sacs. During the process of metamorphosis from tadpole to frog, the animal stops feeding, the horny beak and teeth are lost, the tongue increases in size, the intestine shortens, the gills are lost, the legs grow and the tail is resorbed. The change from a swimming tadpole to a hopping and jumping frog requires the development of a bony skeleton. As far as this is concerned, the animal is an isolated system, hence all the calcium and phosphate for bone formation must be obtained from within the animal itself. As calcification proceeds the calcareous deposits of the endolym-

107

phatic sacs are reduced, in fact they supply all the calcium (fig. 14.3). The phosphate is liberated as a result of the resorption of the swimming tail. From these two sources the animal receives both calcium and phosphate for the production of the bone mineral.

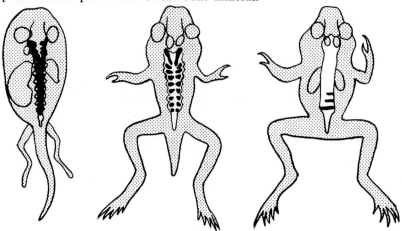

Fig. 14.3. Metamorphosing frogs showing progressive reduction of calcium carbonate deposits (black) within the endolymphatic sacs (white) (after Simkiss).

In the embryos of reptiles, the endolymphatic sac is also filled with calcium carbonate deposits, but this is resorbed during development. The embryos of both birds and mammals also possess these deposits but these too are resorbed. In some female tropical lizards, the endolymphatic sacs form pockets beneath the skin in the neck region and have given rise to the notion among local people that these particular lizards carry their eggs behind their ears. During the egg-laying period these sacs become gorged with calcareous deposits, which disappear when the lizards are no longer breeding. This type of arrangement utilizes the amphibian system to provide at the same time for the manufacturing of a calcareous eggshell. The next stage is that this storage of calcium is transferred to the bones themselves. This is best seen in the evolution of medullary bone in the birds.

One of the crucial problems in the transition from water to land is the difference in the availability of calcium, and the storage in the endolymphatic sacs seems to be the way in which the vertebrates first overcame this particular problem. It has been shown experimentally that in the living frogs or their tadpoles the deposits of calcium carbonate can be readily mobilized, so that they are available physiologically. Of particular interest however is the fact that when every speck has been resorbed and the animals are placed in an environment where they need to resorb still more, the calcareous otoliths are never dissolved. These are never available for these general needs.

PART III

THE SKELETAL SYSTEM

CHAPTER 15

bone as a structural material

15.1. *Strength of bone*

Bone is composed of two contrasting materials, the mineral apatite, which is resistant to compressive forces, and the fibrous protein collagen, which has great tensile strength. Because of this, it is assumed that when force is applied to bone, tensile stresses are taken up by collagen and compressive stresses by the apatite. The tensile strength of a material is recognized as the stress (i.e. force per unit area) which is measured in newtons per square metre (N m^{-2}, a unit also called the pascal, Pa)*, at which the material fails when being pulled apart, the compressive strength the stress at which it collapses under pressure. When these forces are applied the material may yield elastically, that is when the force is removed it will spring back to its original shape. Beyond a certain stress the material will be permanently deformed; it will in fact flow. It is then said to have suffered plastic deformation. The final stage is where the material fails. Substances such as rubber are tough, others such as glass are brittle. One of the most useful descriptions of structural materials is their degree of stiffness. The index of stiffness is Young's modulus of elasticity, often written as E, calculated as

$$E = \frac{\text{stress } (= \text{force per unit area})}{\text{strain } (= \text{elongation per unit length})}$$

Hence rubber has a low modulus of elasticity, collagen a high modulus and glass a huge one.

Bone is an unusual structural material in that its tensile strength is relatively high compared to its compressive strength: 105×10^6 N m^{-2} as opposed to 172×10^6 N m^{-2} (Pa). This is particularly unusual, because bone appears to be a rather brittle material. Bone has been considered as a compound material in the same way as reinforced concrete. The tensile strength of bone is incredibly high when compared with other materials with a high mineral content. For example high duty porcelain has a compressive strength of 552×10^6 N m^{-2}, but a tensile strength of only 55×10^6 N m^{-2}. Collagen has a tensile strength of some 552×10^6 N m^{-2}. The modulus of elasticity (E) of collagen has been calculated as 1.24×10^9 N m^{-2}, that for fluorapatite (hydroxyapatite has not been calculated) along the axis of the crystal comes to 166×10^9 N m^{-2}. If the value for E is accepted for the two components of bone, it is possible to

---

* 700 Nm^{-2} (Pa) approximately equal one pound weight per square inch (p.s.i.).

calculate the relative load borne by each component. According to Currey, at the ultimate stress in tension (103×10^6 N m^{-2}) the apatite will bear 206×10^6 N m^{-2}, the collagen only $1 \cdot 50 \times 10^6$ N m^{-2}. With regard to compression the apatite will have to take 342×10^6 N m^{-2} and the collagen merely $3 \cdot 55 \times 10^6$ N m^{-2}. These results appertain if the analogy of a compound material such as reinforced concrete is correct. Unfortunately it is inconceivable that apatite, a brittle material, could resist a tensional stress of over 190×10^6 N m^{-2}. This being so, it is clear that the reinforced concrete analogy is not appropriate.

Currey put forward a further analogy, comparing the structure of bone to that of fibreglass which is made up of fine glass fibres embedded in an epoxy resin, and is stronger than either epoxy resin or glass. The failure of crystalline or brittle materials under tension is due to minute dislocations in the crystal lattices. These flaws act as local stress concentrators, which open up under tensional forces and thus spread through the material. Any such crack would be forced together under compressional forces and so these do not lead to failure. A developing crack builds up a region of high energy at its front and as a result spreads rapidly. The dislocations in crystal lattices and the small cracks that develop therefrom were first described by A. A. Griffith, from whom they have gained the name Griffith cracks. If the brittle material is in small units and embedded in a substance with a low modulus of elasticity, the energy of the crack front will be taken up as a deformational strain in the embedding medium. In this way the crack will no longer be able to spread through the structural material. This is essentially how fibreglass is organized. The glass fibres have a high modulus of elasticity, the epoxy resin a low one. Such two-phase materials, as they are called, have a modulus of elasticity midway between the two components, but have a greater strength than material composed of only one. The crystalline material, if organized into small units, will thus possess a greater strength in tension than would be possible were it present in bulk. This then is basically the situation in bone. Small crystallites of apatite are associated with an organic matrix of collagen and glycosaminoglycans. As with fibreglass, if the apatite crystallites are aligned along the lines of stress, this will further increase the strength of the material.

With Haversian systems in the compact cortex of limb bones, the alternate arrangement of the collagen fibrils in successive lamellae, which produce a ply structure, further increases the strength of the bone and the ability of the limb to withstand the forces that act upon it. The fibreglass analogy proposed by Currey has greatly improved our understanding of the reasons for the observed strength of bone. R. M. Alexander has, however, pointed out that the strength of fibreglass depends in large measure on the glass fibres being long; if the fibres are less than 10 mm in length the fibreglass is not much use. In bone the apatite crystallites are minute. He therefore suggests that filled rubber provides a closer analogy. This is the substance of which the tyres of motor vehicles and pedal cycles are constructed; carbon black, which is fine soot, is mixed with rubber; the particles are spherical and are of comparable dimensions to apatite crystallites. As with bone, the matrix and particles become firmly attached to one another. The modulus of elasticity is much higher than that of rubber and the tensile strength is also increased. Filled rubber with 50 per cent carbon black can be as much as sixteen times as strong as pure rubber. This analogy is a modification of the fibreglass theory.

In the most recent explanation of the strength of bone, it has been suggested that as yet there are no man-made materials that are constructed on the same plan as bone; the fundamental difference is that the minute crystallites are packed closely together. As Currey recognized, the strength in tension is due to the mineral phase being in small discrete units instead of great big chunks, but, unlike fibreglass and filled rubber, the mineral phase of bone is not embedded in a flabby matrix. The crystallites enclose the organic matrix, there may well be some sort of glycosaminoglycan glue, but basically the crystallites are packed together in close juxtaposition. Griffith cracks do not spread through the material owing to a phenomenon known as the 'Cook–Gordon crack stopper' effect. When the brittle material is in discrete units, if a crack develops and runs through one unit, when it reaches the boundary of an adjacent one, it will meet an interface. In fact the energy of the crack front will probably have opened up a further crack at the interface before the original crack arrives. On arrival there will be a T-shaped crack (Fig. 15.1). The energy of the original crack will be dispersed along the interface. In any event it is difficult to extend a crack by applying a force at right angles to it. This theory explains the observed strength of bone and at the same time approximates to the observed ultrastructural organization of the tissue.

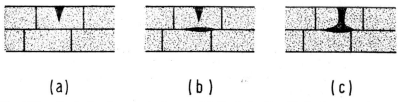

(a)   (b)   (c)

Fig. 15.1. Cook–Gordon crack stopper. (a) Crack begins, (b) further crack opens up in front at interface with adjoining crystal, (c) first crack reaches interface to produce a T-shaped crack, which will only extend with difficulty (after Gordon).

15.2 *Stress concentrations in bone*

Generally speaking one of the most dangerous features of any structural material is the possession of holes or spaces within it, as these act as stress concentrators. Bone is penetrated by blood vessels as well as being permeated by osteocyte lacunae. The channels for blood vessels run in the main along the major axis of the bones and will therefore have little or no stress-concentrating effect. The lacunae of osteocytes are described as being oblate spheroids, that is squashed spheres. Their stress-concentration effect depends on their orientation. In fact they are arranged so that their short axes are normal to the main axis of the bone, thus minimizing the stress-concentrating effect.

Currey has examined this situation and concluded that the osteocyte lacunae actually contribute to the strength of bone by acting as crack stoppers. The stress concentrations at the point of a spreading crack are very large, but if such a crack runs into a lacuna the force will be dissipated. The same principle is employed in the manufacture of plastic rainwear, where round holes border the slits to disperse stress concentrations and thus lessen the chances of the garment being ripped; similarly, rivetted ships are incomparably stronger than welded. Currey tested his views by hitting sections of bone with a little mallet and then looking at the cracks. Most cracks ended in lacunae. It seems that the Cook–Gordon crack stopper effect applies at this level of bone organization as well as at the ultrastructural level.

15.3. *Piezoelectricity in bone*

One of the most significant features of bone as a structural material is its ability to construct itself in response to stress. It has been recognized for many years that bone tends to grow in the direction of a functional pressure, and that the amount of increase in the mass of the bone is proportional to the amount of force applied. It is not well understood how this process actually works. C. A. L. Bassett has propounded the view that such reconstruction is mediated by differences in electric potential. This theory stems from his discovery that bone is piezoelectric, that is when it is mechanically deformed it generates an electric potential. Collagen by itself is known to possess this property of producing a transient electrical polarity, as do glycosaminoglycans. Furthermore the collagen–apatite interface seems to be a semiconductor in that there is an abundance of available electrons in the collagen and a dearth in the apatite crystallites. Any bending at this junction would produce an electric potential. Hence, there are at least three possible ways in which an electrical current could be generated. At present it is not known which of these factors is the most important. Some authors consider that the collagen is the major source, but if the bone mineral is removed from bone the piezoelectric effect is considerably reduced, which suggests that it cannot be due to the collagen alone.

From experiments with pieces of bone of different sizes, Bassett established that the electric potential generated was proportional to the force applied. On bending, the convex side would be under tension and

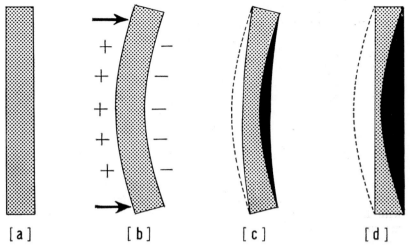

Fig. 15.2. Piezoelectricity in bone (a) piece of bone, (b) when deformed by bending (large arrows) there is developed a positive charge (+) in the region
subject to tension and a negative charge (—) in areas under compression,
(c) this potential difference results in bone resorption in the area of positive
charge and bone deposition in that of negative charge, (d) finally equilibrium
is restored (after Bassett).

[a]   [b]   [c]   [d]

the concave under compression. It was found that a negative charge
was built up on the concave side and a positive charge on the convex
(Fig. 15.2). When a battery was implanted in the legs of dogs so that
two electrodes projected into the bone marrow, it was found that there
was a substantial deposition of new bone around the negative electrode.
From this it was concluded that areas of bone under compression would
by virtue of their negative charge induce the formation of new bone, and
that at the same time bone resorption would occur in the areas subject
to tensional forces. In the experiments on dogs there was no erosion
around the positive electrode, which indicates that there is a more complex relationship between a positive charge and the activity of osteoclasts.

Yet another aspect of electrical effects in the formation of hard tissues
was demonstrated in a study of dissolved collagen. Fibrils came out of
solution oriented at right angles to the direction of an electric field. They
would form more rapidly if the electric current was intermittent. Since
the direction of stress is that of the electric field, this means that the
initial alignment of the collagen fibrils is likely to be at right angles to
the major force to which the developing bone is subjected. This seems
to be the situation in fact, but whether or not this is due simply to
electrical effects is not yet established.

According to Currey, the model proposed by Bassett to account for
the mechanism of bone remodelling under stress will only work for
solid bones, whereas in practice limb bones are cylinders. When a cylinder
is bent the cortex on the outer end of the radius of the bend will be in

net tension and that on the inner end under net compression. This means that both the outer and the inner portions of the cortex will possess the same electrical charge although the amount of the charge will be less at the inner surfaces. On Bassett's model all that would happen is that the concave portion of the bone would become thicker and the convex thinner (fig. 15.3*a,b*). As a result the part of the cortex under tension would become thin and that under compression thick, thus giving the cross-section of the bone a considerable asymmetry. This does not happen, and for this reason Currey rejects the Bassett model of bone remodelling. It seems likely however that the electrical effects demonstrated by Bassett are important in understanding bone re-modelling but that the situation is perhaps more complicated than has so far been suggested. Perhaps this matter could be resolved if Bassett's measuring of the electric potential of bone under stress could be repeated with hollow bone and the charges on the inner and outer surfaces measured. This would either confirm or refute the objections put forward by Currey.

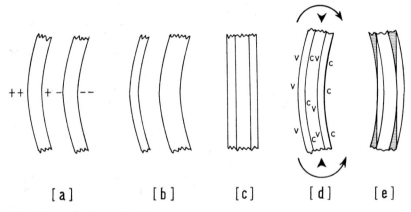

Fig. 15.3. (*a–b*) Currey's objection to Bassett's model of bone remodelling, (*c–e*) Frost's model of bone remodelling; C, concave; V, convex (after Currey).

15.4. *Adaptive remodelling*

Apart from Bassett's model there are two further ones that have been proposed by Frost and Currey respectively. According to Frost, if bone is subjected to either tension or compression above a certain threshold a cycle of cellular activity will be set in motion. First osteoclasts will resorb bone and this phase will be followed by new bone formation. These two phases will balance each other but if the bone is subject to a bending load as well, then osteoblast or osteoclast activity will be inhibited. In the convex parts, the activity of osteoblasts will be inhibited, whilst in the concave parts osteoclastic activity will be reduced. Differential bone deposition and resorption will take place until the bone achieves the shape it had before the bending load was applied (fig. 15.3c,d,e). When the bone is no longer subjected to the compressive bending load, the bone will be seen to have a curvature opposite to that of the original bending, so that with subsequent bending the bone will become straight again. The model of Frost's seems to be applicable in all normal circumstances that are encountered. It will not work if ever a bone is loaded in net tension—indeed it is always assumed that a bone is never loaded in this way.

Currey has questioned this assumption, as there are bones, or at least important parts of bones, that are habitually subject to tensional forces. The projecting processes at the points where ligaments attach, such as the acromion process, the calcaneum, the coronoid process of the lower jaw, and the angular process of the jaws of carnivores, are all subject to tensional forces when in action; the fibula of jumping rodents is also loaded in tension during the jump. The forelimbs of brachiating primates swinging through the trees must be loaded in tension, as must be all the limbs of sloths and bats roosting.

On the models of both Bassett and Frost such bones should disappear by being completely resorbed, but this they do not do. Currey has proposed a further model which will account for the existence of these bones in tension as well as the more normal under compression. In this he suggested that the cells have some means of recognizing whether the net stress on the bone was either compressive or tensional. With the application of a local stress, the reaction induced would be to re-establish the original equilibrium. Hence, if a bone normally under compression is subjected to local further compression, bone will be deposited, and if subject to tension bone will be eroded. In contrast to this situation, a bone under net tension if subjected to further tension will have further bone added whereas compressive forces will lead to bone resorption. It is not known how osteocytes could accomplish the necessary measuring of the strain but their shape is ideal for determining changes in local strain over short distances. There is the possibility that changes in electric potential consequent upon strain could be responsible. Changes of electric potential produced by strain would not be of any value in determining the net tensile or compressive stress and Currey suggests that this may be monitored by nerves within the osteones.

117

CHAPTER 16

biomechanics of joints

16.1. *Types of joint*

The separate units of the vertebrate skeleton are, with few exceptions (such as the penis bone of carnivores, cetaceans, rodents etc.), in contact with one another. The nature of the connection of the different parts varies from one part of the skeleton to the next. There are four basic types of joint. Where bone is joined to bone by means of collagen fibres, the joints are known as syndesmoses or fibrous joints. Examples of this type occur in the sutures between skull bones. A different type of syndesmosis that allows a limited degree of movement is found in the attachment of the teeth to the bone of their sockets. Sharpey's fibres, which are bundles of collagen fibres, inserted in the cementum of the roots of the teeth, form a sling which is incorporated into the alveolar bone of the jaws. When bone ends up fused directly to bone this junction is known as a synostosis and in this instance no movement whatsoever is possible.

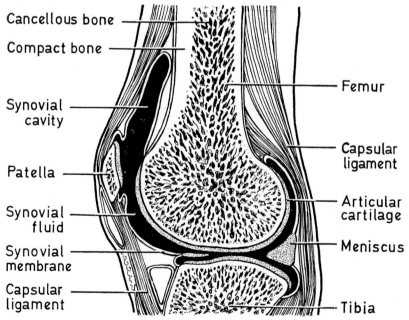

Fig. 16.1. Synovial joint (knee) (after Ham).

Joints that allow a much greater degree of freedom are found in symphyses, where the surfaces of the bones are united by fibrocartilage. They are sometimes known as cartilaginous joints. They are present between vertebrae and in the pubic and mandibular symphyses. In many mammals movement at the mandibular symphysis plays an important part in the functioning of the jaws. The most familiar joints, however, are the synovial, in which the junction is enclosed in a capsular ligament within which there is a lubricant, the synovial fluid (Fig. 16.1). These joints allow rapid movements to be made, and at the same time are capable of bearing considerable loads. This chapter is concerned with the biomechanics of this last type.

16.2. *Engineering principles*

In the consideration of moving parts, be they in machinery or the skeleton, there are two aspects; one is the nature of the bearing material and the other that of the lubricant. When two objects are brought together with the intention that they will move the one relative to the other, it is necessary for there to be some kind of lubricant between them as otherwise the friction between them will cause wear and damage. No matter how smooth the surfaces, there will always be minute irregularities, known as high spots or asperities. In order to prevent these high spots from coming into contact and generating heat from the resultant friction, it is necessary to have a lubricant that will as far as possible keep the two surfaces apart. In practice the asperities still come into contact during movement. This type of lubrication is termed boundary-layer lubrication, and is found in slow-moving parts (Fig. 16.2a). The lubricant ideally adheres to the articulating surfaces, and in machinery consists of greases with a high viscosity. In the skeleton the nearest approach to this type of lubrication is the slow-moving joints such as the neck. If the neck is moved slowly a faint grating can be heard which is due to asperities grinding on each other.

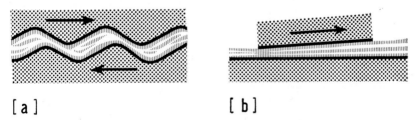

[a]　　　　　　　　**[b]**

Fig. 16.2. (*a*) Boundary layer lubrication, (*b*) hydrodynamic lubrication (after Swanson and Freeman).

With fast-moving joints the situation is very different. With a thick lubricant only slow movement is possible; the thinner the layer of lubricant the faster can become the movement. In joints of this type the lubricant is forced between the moving parts and the lubrication is termed hydrodynamic (fig. 16.2b). At the same time the articulating surfaces are covered with a bearing material. The nature of the bearing material is critical, since to be an effective bearing material it must have a number of special properties. It must have a smooth surface in which the asperities or protuberances are less than the thickness of the layer of lubricant. The surface should allow the lubricant to adhere, and if the bearing material is capable of absorbing the lubricant this will be a further advantage. Finally the bearing material should be able to withstand shear as otherwise there is always the possibility that it will be pulled off. In machinery, especially motor vehicles, the bearing materials are soft alloys of 'white metal' named babbitts, after Isaac Babbitt, who invented them in 1839. A typical bearing comprises 80 per cent tin, 8 per cent antimony and 4 per cent copper; however, the composition is extremely variable and these alloys range from 90 per cent tin and no lead to 80 per cent lead and only 5 per cent tin. In the big end of a motor car the connecting rod articulates with the crankshaft. The articulating surfaces are covered with white metal, and the lubricant is pumped between them from the sump. If the vehicle is run without sufficient lubricant, the friction and consequent heat tears the bearing surfaces, a situation which is paralleled in the skeleton, where it is termed osteoarthritis.

16.3. *Synovial joints*

All the features and requirements in the fast-moving parts of machinery can be paralleled in synovial joints, but as well as being concerned with rapid movement, these joints have to be capable of load-bearing. To deal first with the question of movement, the articular surfaces are covered with a bearing material, the articular cartilage (Fig. 16.1). The surface of the articular cartilage is remarkably smooth. In electron micrographs projections of collagen fibres can be observed but their extent is less than the thickness of the synovial fluid so that no wear occurs. The synovial fluid adheres firmly to the articular cartilage, which has the further ability to soak up liquid from the fluid, which can be subsequently squeezed out again. Because of this facility, synovial joints have been considered comparable to weeping joints.

The shear strength of the articular cartilage is given by the collagen fibres that run between the columns of chondrocytes arranged normal to the articular surface. At this surface, the collagen fibrils run tangentially and it seems likely that the collagen fibres form arcades. The cementing matrix of chondroitin sulphates is elastic but weak, whereas the collagen is less elastic but strong in tension. This combination produces a surface that can be deformed but which readily reverts to its original shape. But more important, the arrangement of the collagen arcades renders it extremely difficult for the articular cartilage to be plucked off. The base of the articular cartilage is calcified in an irregular jagged zone beyond which the trabeculae of the spongy cancellous bone run. This region of calcified cartilage further assists in anchoring the articular cartilage, thus helping to resist shear.

From the standpoint of the engineer, the articular cartilage fulfils all the requirements of the best bearing materials. One of the important features of a fast-moving joint is that the two surfaces should be closely adpressed. If the bones alone are examined in the limbs of vertebrates many of the joints would appear to have large spaces in them. In fact there are never any such spaces. The gaps are filled by pads of fibro-cartilage, the menisci, which in the knee-joint are the semi-lunar cartilages. These structures ensure that the layer of synovial fluid remains thin and they also assist in forcing the fluid through the joint so that any heat generated is carried away.

16.4. *Synovial fluid*

In all moving systems the properties of the lubricant are of crucial importance. It is no use having the most efficient and effective bearing surfaces if the lubricant is inadequate. In fact synovial fluid is one of the most remarkable substances in the body. When a drop of liquid is placed on an optically flat surface and a lens is pressed down on top, with water or blood plasma the two objects come in contact readily; but when synovial fluid is used the two objects can never be brought into contact. When the pressure is released the lens bounces up. Synovial fluid is seen to be elastic. This property can be readily demonstrated by swirling synovial fluid in a glass. When this action is stopped suddenly the fluid swirls back in the opposite direction. This property is due to the glyco-saminoglycan hyaluronic acid. According to C. F. Phelps, in 100 cm^3 of a 0.1 per cent solution of this material 'our molecular gnome would have to walk over ten million kilometres to inspect all the chains' of the molecules. It has been further estimated that in synovial fluid 1 g of hyaluronic acid occupies 12 litres of solvent. This substance has enormous viscosity, but if this were all it would be singularly inappropriate as a lubricant. However, it possesses what is termed anomalous viscosity or thixotropy; the viscosity alters, depending on the degree of shear to which it is subjected. With low shear the polysaccharide chains are all tangled up so that the material has a high viscosity; with a high degree of shear the chains line up with the flow lines and the viscosity collapses. This behaviour is functionally exceedingly important. With a slow deliberate movement entailing the precise positioning of a limb, the synovial fluid would be almost solid in the joint space. In contrast, with a very rapid movement the synovial fluid would be a thin liquid. Phelps uses the example of a soldier crossing a minefield in contrast to a footballer playing a game. The requirements of the two men and their legs are very different but the anomalous viscosity of hyaluronic acid serves both with equal effectiveness.

The nature of synovial fluid enables it to play a further necessary role. Synovial joints have to bear load. Under pressure the synovial fluid forms a firm gel, which ensures that the two bearing surfaces are never brought into direct contact. When the pressure is released the system is restored. The small patches of hyaluronic acid that form a solid gel during loading bear the greater part of the animal's weight (Fig. 16.3). Take for example the situation of a man standing on one leg (a feat that is quite feasible). The weight is then transmitted through the articular cartilage to the fine tracery of trabeculae at the head of the long bone, through which it is dispersed to the compact cortex of the shaft. The fine tracery is so arranged that the load can be transmitted to the shaft from any part of the articular surface.

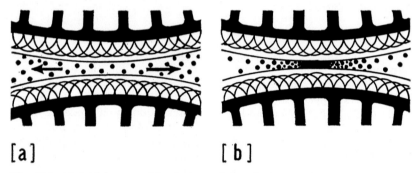

[a]　　　　　　　　　[b]

Fig. 16.3. (a) Initial stage of loading at a joint, (b) final stage with formation of solid gel of hyaluronic acid (after Swanson and Freeman).

The dual functions of joints—movement and weight bearing—are accomplished by the special qualities of both the bearing material and the lubricant. Both of these are organized on the most efficient of engineering principles.

CHAPTER 17
mechanics of mammalian mastication

17.1. *Origin of the mammalian masticatory apparatus*

The hardest and most resistant of vertebrate hard tissues are found in the teeth. Primitive teeth were concerned merely to prevent the escape of prey, later in evolution they came to play a part in the preparation of the food for digestion by either crushing it or chopping it up. Basically teeth perform two functions, the ingestion of food and its subsequent mastication. The transition from the food-trapping function to one of mastication can be traced in the evolution of the paramammals, the mammal-like reptiles, among which can be found all gradations from a reptilian grade to the mammalian.

From an examination of the jaws it is possible to work out the changes that must have taken place in the evolution of the jaw musculature to accomplish this. In fact any discussion on the functioning of the dentition must be in the context of the jaw apparatus. With the most primitive of the paramammals, the pelycosaurs, the dentition consisted of numerous sharp-pointed teeth which served to prevent the escape of fish. These paramammals were semi-aquatic and inhabited swamps and lakes in Carboniferous and Permian times. The articular bone of the lower jaw articulated with the quadrate bone of the skull. The jaw was closed by means of a muscle, the capiti mandibularis, which originated on the skull medial to the temporal opening and inserted on the inner surface of the lower jaw behind the teeth but in front of the joint. With food held in the jaws the force of the capiti mandibularis muscle would exert an upward force on the food, which would exert a downward force on the jaw. However, the lines of action of these two forces are not in line and hence the jaw will also be pulled up at the articulation. This means that the skull will exert a downward reaction on the jaw. Any bite will thus transmit an upward force at the jaw articulation (fig. 17.1 *a*).

From the early semi-aquatic paramammals there arose fully terrestrial carnivorous forms such as *Dimetrodon* that had developed large canine-like stabbing teeth. The shape of the lower jaw had altered (fig. 17.1*b*). There was a development of the coronoid and the angular bone. Since the shape of a bone reflects to a certain extent the organization of its associated musculature, it is reasonable to infer that at this stage the simple jaw-closing muscle block had divided into two separate components, one inserted on to the developing coronoid process which not

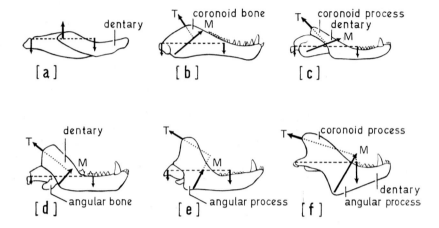

Fig. 17.1. Evolution of mammalian jaw. (*a*) Primitive semiaquatic paramammal with undifferentiated jaw musculature, (*b*) carnivorous primitive paramammal, *Dimetrodon,* showing division of muscle into two components, temporalis (T) and masseter (M), (*c–e*) advanced paramammals showing gradual approach to mammalian condition, (*f*) primitive Triassic mammal, *Diarthrognathus* (after Crompton).

only closed the jaw but pulled it backwards and another inserted on the angular which besides closing pulled forwards. The former can be identified as the precursor of the temporalis and the latter of the superficial masseter. The deep masseter, which has an almost vertical action, must also have been present, as well as the more anterior part of the temporalis, which similarly acts vertically. The intersection of the main lines of action of the temporalis and superficial masseter is situated somewhat in advance of that of the old reptilian capiti mandibularis. This means that the force exerted on the jaw by food would be partially taken up by this set of synergistic muscles. By the same token the downward reaction through the articulation of the jaw would be perceptibly reduced.

It is possible to trace a series of paramammals which show a progressive increase in the development of the coronoid process and also the angular. With these changes there is an increase in the size of the dentary bone. Eventually the coronoid and angular processes of the dentary develop to replace the bones of the same name (fig. 17.1*c–e*). The bones of the posterior part of the lower jaw become progressively reduced, in particular the actual bones of the joint, the quadrate and the articular. In the most advanced paramammals, the therapsids, the intersection of the two lines of action of the jaw muscles is in line with the posterior teeth. The force on the jaw from the food will thus be entirely met by the musculature. At this stage there will be no downward reaction at the joint itself. This explains the apparently anomalous situation, where the jaw musculature was becoming more massive and powerful, while the ability to bite with great strength was coupled with the virtual disappearance of the bones of the jaw articulation. In fact the quadrate and articular become incorporated into the mammalian middle ear as additional sound amplifiers. Several advanced forms are known in which the dentary bone has come into contact with the squamosal or temporal bone of the skull, while still retaining the old quadrate-articular joint (fig. 17.1f). By definition an animal with a squamosal-dentary joint is deemed to be a mammal, one with the quadrate-articular a reptile. In these instances the consensus is that these borderline forms are accepted as mammals.

The first mammals (by definition) appear at the end of the Triassic period, with the ability to give their prey an effective and presumably fatal bite. At the same time the looseness of the jaw articulation would have allowed a greater freedom of movement than that found in any other contemporary vertebrates, thus allowing the animals to chew their food for the first time.

17.2. *Muscles of mastication*

In the paramammals it is possible to infer the general arrangement of the jaw musculature but its details cannot be determined; this can only be accomplished by examining living primitive mammals. In essence it is useful to consider the mandible as supported from the skull by two muscular slings. The temporalis muscle, which originates from the squamosal and parietal, inserts on the coronoid process of the dentary and holds the jaw from above (fig. 17.2). The masseter muscles run

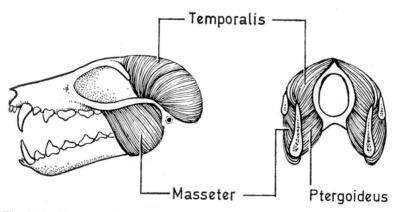

Fig. 17.2. Diagram of the jaw musculature of a mammal (fruit bat) showing the upper sling formed by the temporalis and the lower by the masseter and pterygoideus.

125

from the outer surface of the posterior part of the jaw to insert on the zygomatic arch or further forward. On the internal or medial surface the pterygoid muscles run from the braincase. The masseter and pterygoid muscles in conjunction act as a sling around the lower end of the jaw suspending it from the skull (fig. 17.2).

The muscles of the jaw apparatus have to subserve two distinct functions. The most obvious is to move the jaws, but at the same time these have to be held firm even when the animal is not eating. The muscles that stabilize the jaw apparatus, the holding muscles, are known as tonic in contrast to the actively moving or phasic muscles. Muscles are made up of two types of fibre that can be distinguished histologically by staining with Sudan Black B. The tonic fibres take up a dark stain, the phasic appear white. By examining the relative proportions of the two types of fibres, it is possible to determine whether a particular muscle or part of a muscle is concerned mainly with holding or active moving.

Such histological studies complement the conclusions that can be made from an examination of the anatomy of the muscles. The temporalis muscle has two main parts; there is a posterior which has a mainly horizontal action and an anterior which exerts a more vertical. The vertical component has a high proportion of tonic fibres, the horizontal virtually none. The masseter muscle is divided into two, a deep masseter which has a vertical line of action, originating on the zygomatic arch, and with a high proportion of tonic fibres, and the superficial masseter with few tonic fibres and a more horizontal line of action. It is evident that the superficial masseter and the posterior temporalis, which are antagonistic to one another, are responsible for pulling the jaw forwards and upwards and backwards and upwards respectively. The deep masseter and anterior temporalis are primarily concerned with holding the jaws.

The temporalis and masseter muscles are concerned only with movement in vertical and fore-and-aft directions. Lateral movements are the concern of the pterygoid muscles, which originate from the base of the skull and insert on the inner surface of the mandible. There are two parts of the pterygoid, the external (or lateral) which attaches near the articulation, and the internal (or medial) which runs from the angle of the jaw anteriorly forwards. The external pterygoid has many tonic fibres and is concerned with controlling movement at the joint. The internal pterygoid is phasic, and as well as pulling the jaw forwards in concert with the superficial masseter, also pulls towards the midline. Lateral movements of the jaw are hence the province of the internal pterygoid. As well as the muscles already described, there is a further set that attach on the anterior part of the inner surface of the mandible—the mylohyoid and digastric. These pull the jaw backwards and downwards. They actively open the mouth and also prevent the jaws being pulled out of their sockets.

126

17.3. *Function of the mammalian masticatory apparatus*

Generally speaking, mammals can only chew on one side at a time, although it is known that rodents are capable of chewing on both sides simultaneously. The chewing cycle comprises three strokes. From a position in which the jaws are open, the lower jaw is lifted both upwards and sideways. This is the preparatory stroke and brings the outer or buccal part of the lower teeth into contact with the outer part of the upper teeth. During the power stroke the lower teeth are sheared across the upper—they are moved both vertically and medially towards the midline. This is the effective chewing action after which the jaw is opened again, the recovery stroke (fig. 17.3).

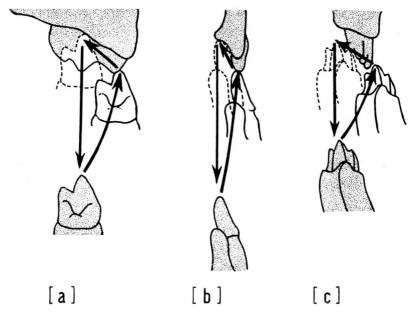

[a]   [b]   [c]

Fig. 17.3. Chewing cycle. Diagram of cross-section through jaws in region of molar teeth, midline towards the left, heavy arrows indicate movement of lower teeth in relation to the upper. (*a*) Insectivore, (*b*) carnivore, (*c*) herbivore (after Crompton and Hiiemae).

In the advanced carnivores such as the cats, the main action of the molar teeth is a scissor-like cutting of the food. The main movement of the jaws is in the vertical plane and the lateral component of the power stroke is minimal (fig. 17.3*b*). The temporalis muscles are strongly developed and, together with the masseter, enable the animal to effect a powerful bite, to slice through flesh or crunch bone. However, before the food can reach the back teeth it has to be stripped off the bone, and this entails a lot of pulling and tugging, the effect of which is to pull the jaws forwards out of their sockets. The digastric, which with the temporalis resists such forces, is very well developed. Although the problem of dislocating the jaw forwards may thus be overcome there is always the possibility of these two muscles over-doing things once the flesh being tugged comes away. The masseter will tend to counteract these forces but in fact the jaw articulation itself is constructed to resist them. The articular condyle of the mandible is in the form of a transverse cylindrical rod which fits tightly into a deep furrow in the squamosal or temporal bone. This articulation allows movement in the vertical plane but severely restricts lateral movement. Furthermore, at the posterior margin of the articular fossa there is a hook-like bony flange which will similarly prevent the jaw being tugged forwards out of its socket.

The opposite extreme of specialization is seen in the herbivores, such as horses, cows and sheep. In these animals the food is ground exceedingly fine in a kind of millstone action. The lateral component of the power stroke is greatly exaggerated and the main chewing action is from side to side (fig. 17.3c). This is familiar to anyone who has seen a cow or sheep chewing the cud. There is no need for a large gape to ingest the food and in fact the herbivores open their mouths but slightly. At the same time there is no need for a powerful bite. The temporalis is poorly developed, and unlike the condition in the carnivores the coronoid process is very small. In contrast, the angle of the jaw is expanded to give a large area of muscle attachment. The internal pterygoid is greatly developed, as is the masseter. Both these muscles, in concert, are responsible for the lateral movements of the jaws, the masseter and internal pterygoid both having a vertical component. The digastric is important in pulling the jaw backwards.

The condyle of the mandible in the herbivores is flat or faintly concave and it articulates with an ill-defined flat area on the squamosal which can hardly be designated a fossa. The articulation is exceedingly loose, enabling the jaws to slide about with considerable ease to grind fine fibrous vegetable matter such as grass.

The third specialized type of jaw is that of the gnawers and nibblers, the rodents and lagomorphs (rabbits and hares). A cursory view of the jaw articulation shows a narrow condyle elongated antero-posteriorly which fits into a similarly elongated groove. This has led to the general notion that the main action of rodent jaws is a to-and-fro grinding movement. From a study of slow motion X-ray cine-films, it has been shown that this is not so.

The jaw apparatus of rodents is adapted to perform two quite distinct tasks. First of all, food is obtained by the chiselling action of the incisors. These are continuously growing teeth with the enamel present only on the outer, buccal, surface. Differential wear ensures an eversharp cutting edge. Rodents maintain the cutting surfaces of their incisor teeth by sharpening them against each other, an activity known as thegosis. For gnawing and nibbling the lower jaw is drawn forward so that the incisors can be brought into direct contact, into occlusal relationship. There is a large gap, the diastema, between the incisors and the cheek teeth in which the food is collected. It is possible for rodents to close off the incisors from the mouth so that they can chisel, say wood, without its entering the mouth only to have to be spat out. When a rodent is gnawing, the lower cheek teeth are not in any sort of functional relationship with the upper teeth.

After gnawing, the mandible is retracted so that the cheek teeth are in alignment with the upper ones. The food is pounded and sheared into small particles with an upwards and forwards movement towards the midline. The most important muscle in the rodent jaw apparatus is the masseter; this is the main chewing muscle and the superficial masseter has a horizontal part which acts in pulling the jaw directly forwards. The temporalis and digastric retract the jaw. Contrary to expectation, the internal pterygoid is well developed confirming that there is an important lateral element in the chewing action.

From an examination of the shapes of the mandibles and the nature of the articulation, it is possible to work out the general functioning of the jaw apparatus. This is only worthwhile if the musculature is similarly studied. The rodents provide something of a cautionary tale in this context. In these animals the entire apparatus with the apparent emphasis on an antero-posterior movement is developed merely for the jaws to achieve two separate and distinct specialized roles.

CHAPTER 18

form and function in teeth

18.1. *Origin of the mammalian molar*

From a dentition concerned merely with preventing food escaping there evolved one which prepared it for digestion. During this evolution of the jaw apparatus, there were changes involving the evolution of the secondary palate which provided a separation between the air and food passages. This ensured that the animal could breath and chew at the same time; it also enabled the young to breathe while suckling.

In the last few years there have been discoveries of fossil teeth that

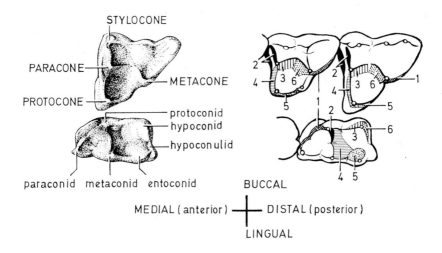

Fig. 18.1. Tribosphenic molars. Upper drawings; occlusal (ventral) view of upper molars; lower drawings : occlusal (dorsal) view of lower molars. Teeth are conventionally illustrated in this way so that the occlusal relationships of the cusps can be readily perceived. Triangle of upper molar (paracone, protocone, metacone) termed trigon; triangle of lower molar (paraconid, protoconid, metaconid) the trigonid; posterior heel (bounded by entoconid, hypoconulid, hypoconid) the talonid. Right hand diagram showing matching shearing facets on occlusal surfaces. Note on dental terminology: Buccal towards the cheek (i.e. external surface); Lingual towards the tongue (i.e. internal surface); Medial towards the midline (i.e. anterior in molars); Distal away from the midline (i.e. posterior in molars). With regard to incisors the buccal surface faces anteriorly whereas in molars it faces laterally, hence the need for a terminology which describes the same surface of a tooth regardless of its position and orientation in the jaw. (After Crompton and Hiiemae.)

enable the history of the tritubercular or tribosphenic molar of the mammals to be traced. It is now possible to show how the primitive triangular molar arose but, what is more important, following the work of K. Hiiemae and A. W. Crompton, the functional reasons for these changes can be understood. The primitive tribosphenic molar is essentially a triangle of cusps; in the upper jaw the apex faces inwards or lingually, in the lower outwards or buccally. These triangles of the occlusal surfaces fit one within the other. However, the lower molars develop a heel or talonid which forms a basin into which the main cusp (the protocone) of the upper molar runs. The relationships of the cusps to one another remains constant. In no instance will a cusp that occludes against one side of a ridge ever change to the opposite side. In exactly the same way the relationships of the shearing facets on the teeth remain constant (fig. 18.1). Although it is possible to trace the history of the different shearing facets, it is easier to follow the evolutionary history of the different cusps.

In the primitive paramammals the teeth formed sharp points and were flattened from side to side bucco-lingually, the long axis of the tooth's cross-section running in the same direction as the length of the jaw. When the jaws were closed the line of the lower teeth was enclosed by that of the upper. The relationship of the upper and lower teeth was alternate, that is each lower tooth occluded against two upper. This type of dentition formed an ideal trap and at the same time could function as a general stabbing and perhaps tearing mechanism.

The advanced paramammals showed a certain improvement in that they developed cusps on their cheek teeth. The cusps were aligned in the direction of the jaw, that is antero-posteriorly or in dental terms medio-distally. The central cusp in the upper tooth is the paracone, the posterior the C cusp and the anterior is the stylocone. In the lower teeth the cusps are from front to back the paraconid, protoconid and metaconid (see fig. 18.2).

The earliest mammal on the evolutionary line leading to the modern forms, recently described and named *Kuehneotherium* by K. A. Kermack, shows the first stages of the development of the tribosphenic molar. The three cusps of the advanced paramammal are still in evidence, but for the first time are no longer in alignment. The paracone is offset towards the lingual side, the C cusp (known only in primitive teeth) and the stylocone are still in line. The tooth shows the beginning of a triangular arrangement of its cusps. The lower teeth have a very high protoconid offset buccally, and paraconid and metaconid. Of special significance is the beginning of a heel or talonid at the posterior (distal) margin of the tooth with a small cusp, the hypoconulid. When the teeth are in occlusion, this cusp acts as a stop to prevent the paracone of the upper tooth impacting food particles into the angle, the embrasure, between the teeth or running down to stick into the gums. With triangular teeth

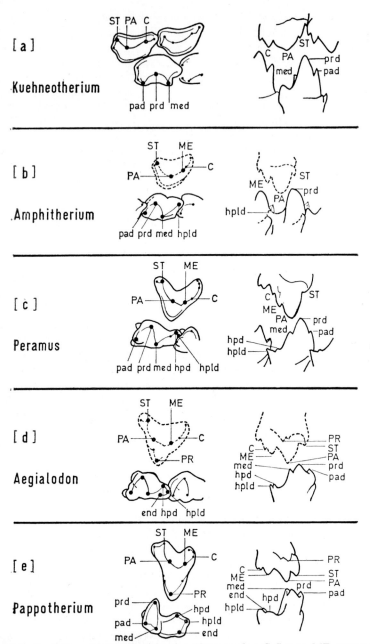

Fig. 18.2. Evolution of tribosphenic molar. Upper molar: C C-cusp, ME metacone, PA paracone, PR protocone, ST stylocone. Lower molar: end entoconid, hpd hypoconid, hpld hypoconulid, med metaconid, pad paraconid, prd protoconid. Orientation of left hand column as in fig. 18.1: right hand column lateral view, anterior to right (after Crompton)

interdigitating, it is of very great importance to have such a stop for the prominent cusps that fit into the angle between the teeth.

The action of the jaws is not simply an up-and-down movement; the lower jaw moves upwards and laterally so that the outer buccal surface of the lower teeth are in contact with the outer buccal edge of the upper. The action of the power stroke of the action of the teeth, when the lower teeth are sheared across the upper in an anteromedial direction, results in a pattern of shear facets developing on the teeth. At the end, and only at the end, of the power stroke does the paracone hit the stop of the heel of the lower tooth. In spite of the emphasis on the evolution of the pattern of cusps, from a functional point of view the ridges joining the cusps may be more important. These ridges shear against one another and a series of wear facets can be recognized. In *Kuehneotherium* there are five sets of wear facets. During the subsequent evolution of mammalian teeth the relationships between cusps and ridges remain constant. It is not possible for a cusp to occlude on the other side of a ridge, as this would mean there would have to be a point where a cusp occluded at the crest of a ridge which simply could not work.

The next stage is found in the Jurassic *Amphitherium* which is known only from its lower teeth (fig. 18.2 *b*). Nevertheless, it is possible to work out what the opposing teeth must have been like. The most obvious development is that in cross-section the teeth are triangular. At the posterior internal (disto-lingual) margin of the lower teeth, there developed for the first time a definite talonid or heel. The talonid slightly overlaps the anterior part of the next molar behind, thus producing complete protection of the gum from having food pushed into it. The hypoconulid is well developed and is joined to the metaconid by a ridge. The paracone shears against the buccal side of the ridge running from the hypoconulid to the metaconid. The metacone now forms a prominent cusp, while the C cusp is greatly reduced. The shear facets on the teeth make up six sets, five as in *Kuehneotherium* with the addition of a further set, and this pattern of facets is retained throughout the subsequent development of the tribosphenic molar.

At the beginning of the Cretaceous is found *Peramus*, which illustrates a further development of the tribosphenic molar (fig. 18.2 *c*). In the upper teeth there is a considerable lingual extension of the crown, with the beginnings of the hint of a new cusp. In the lower teeth the talonid basin shows the development of a new cusp, the hypoconid, on the buccal side of which the paracone sheared. On the lingual side of the hypoconid a talonid basin was formed. The relative positions of the paracone and metacone had changed somewhat so that they were more or less in line, the paracone no longer being set off at the lingual corner of the tooth. The C cusp was further reduced.

Kermack and his co-workers in 1965 described the first known tribosphenic molar from the Lower Cretaceous (fig. 18.2 *d*). Only a single lower molar was found and this was named *Aegialodon dawsoni* in honour of Charles Dawson, who was one of the first people to discover fossil mammals in the Cretaceous rocks of England, but who is now only remembered for his association with the Piltdown Man hoax. The talonid basin was expanded, and the high crowned original triangular tooth the trigonid was reduced in height. A further cusp was developed on the talonid, the entoconid. In *Aegialodon* the full complement of talonid cusps were now present. From the form of this lower molar it is evident that the upper molar must have possessed a well developed cusp on the lingual extremity of the crown. This is the first time the protocone is found in the mammalian molar. The upper molar at this stage of evolution clearly possessed four major cusps. At the end of the power stroke the protocone ended up fitting into the hollow talonid basin. From the stage represented by *Aegialodon* the subsequent evolution of the tribosphenic molar is straightforward, merely entailing the development of features already in existence.

In the later Cretaceous mammal *Pappotherium*, the protocone is more developed but is still much smaller than the paracone or even the metacone. The talonid basin is further expanded. On the ridge linking the protocone with the paracone and metacone two minor cusps develop. By the end of the Cretaceous more typical tribosphenic teeth are in existence. In these the major cusp of the upper molar is the protocone and the original reptilian cusp on the anterior lateral or mesiobuccal side finally disappears as a major cusp. The talonid basin increases until it almost equals the trigon.

In an advanced type of tribosphenic tooth, such as in the primates, a further cusp, the hypocone, develops on the posterior inner or linguodistal edge of the upper molar. The lower molars become more rectangular, the talonid basin and the trigonid becoming sub-equal, with the paraconid finally disappearing.

Among primitive mammals the action of the teeth is a more or less vertical pounding and puncturing action. When the food has been sufficiently broken down, it is then sheared into finer particles, when the lower molars are driven up both forwards and towards the midline. In this way the ridges on the teeth shear against each other and the shear facets are produced. This shearing action can also take place when there is no food between the teeth. This action serves to maintain sharp shearing surfaces on the teeth.

In many mammals, there are arcuate ridges on the roof of the mouth, the palatal rugae. In primitive forms such as the opossum, these ridges run from the protocones of the molars and mark wide grooves or corrugations in the palate. At the end of the shearing power stroke, particles of food are collected into these grooves to form a bolus, which the tongue then passes back to the oesophagus.

134

18.2. *Specializations of the mammalian cheek teeth*

The cheek teeth, that is those borne in the maxilla or, in the lower jaw, biting against these, are in most mammals divided into premolars, of which there are both milk and permanent sets, and molars, of which there is only one set. In some mammals, such as man, they can be distinguished by their form, in others they cannot.

The form of all mammalian cheek teeth can be traced from the basic tribosphenic molar that evolved from such primitive mammals as

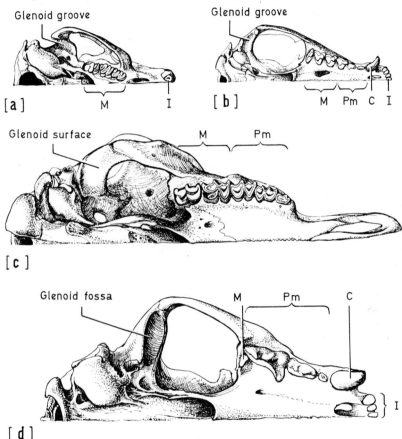

Fig. 18.3. Upper dentition of selected mammalian skulls. (*a*) Rodent (coypu) showing longitudinal glenoid groove for articular condyle of mandible, three molars and single incisor, (*b*) marsupial (opossum) to show primitive unspecialized dentition, (*c*) herbivore (sheep) with flat articular glenoid surface, molarized premolars and absence of upper incisors, (*d*) carnivore (cat) with deep transverse articular glenoid fossa with bony flanges both anteriorly and posteriorly, shearing carnassial (fourth premolar but first molar in the lower jaw) and stabbing canine. I, incisors; C, canine; Pm, premolars; M, molars.

135

Kuehneotherium. The insectivores and bats still retain this basic tooth pattern, which serves well to puncture the hard exoskeletons of insects as well as to chew them up into minute fragments. This type of molar is unfortunately less than adequate when it comes to slicing up large pieces of flesh or grinding fibrous vegetable matter. To illustrate the range of tooth form it is sufficient to consider the two extremes of the advanced carnivores and herbivores.

Among the carnivores, the action of the cheek teeth is scissor-like. The entire force of the bite is concentrated towards the posterior part of the jaw. Two teeth in particular develop into effective cutting blades; these are the carnassial teeth, which are the fourth upper premolar and the first lower molar. The full complement of the mammalian dentition is greatly reduced. In the carnassial teeth the form of the tooth is elongated antero-posteriorly and although the main cusps are still present only a single shearing-ridge is exaggerated (fig. 18.3).

In herbivores the problems are somewhat different. Herbivores need to consume large amounts of plant material to obtain sufficient protein and this involves dealing with tough and often siliceous materials. The occlusal surface of the teeth has to cope with this and the problem is compounded by the fact that mammals possess a limited number of teeth during their lifetime, in contrast to reptiles and fish which have a continual succession of teeth. Mammals have only one and a half generations of teeth; there are two sets of incisors, canines and premolars but only one set of molars.

This limitation is overcome to some extent by an increase in the height of the teeth, which are then termed hypsodont. These teeth have open roots and are continuously growing, so that as wear proceeds, the teeth are able to keep pace.

The herbivore dentition shows no reduction of the cheek teeth. The premolars that in primitive mammals are generally simpler in form than the molars are hardly distinguishable in the herbivores, being 'molarized'. The most obvious feature is the apparent suppression of the cusps. At the same time the shearing ridges connecting the cusps become accentuated and complicated folds develop on them. In cows and sheep, as well as horses, these ridges on the teeth are aligned parallel to the tooth row. Since the action of the jaws has a major lateral component, these ridges run normal to this and hence present a maximum surface as the lower teeth grind across the surface of the upper.

In herbivore cheek teeth the crowns are made of enamel, dentine and cementum. Once the teeth have erupted and are in use the differential wear of these three tissues always ensures that the sharpness of the enamel ridges will be maintained. These parallel shearing ridges effectively grind plant material and break down the cellulose cell walls. With rodents, the convolutions of the enamel ridges are frequently aligned transversely suggesting that the grinding action may have an

136

anterior component. Teeth with an occlusal surface of ridges are termed lophodont; where ridges have a crescent shape the teeth are known as selenodont. There are further methods of increasing the effective working surface of the teeth of herbivores. This is by developing numerous accessory cusps, as for example among the pigs and bears, which gives the crown a bubbly appearance, a type of tooth termed bunodont.

Mammalian teeth are the most variable in form of any class of vertebrate and closely reflect the animal's feeding habits and diet. Among the dolphins and toothed whales, the teeth have become secondarily simplified to sharp single cusped points, while their numbers have also been enormously increased. The dentition is thus exactly comparable to that of the primitive paramammals, and served similarly to trap food and prevent its escape. In mammals such as the edentates, that have taken to a diet of ants and termites, the dentition is also secondarily simplified and in some cases entirely lost. Finally with man today one of the main functions of teeth is cosmetic. The poor state of teeth in civilized societies led A. P. Bystrow to portray man of the future as being entirely toothless—a daunting prospect.

CHAPTER 19

static and dynamic skeleton

19.1. *Weight bearing*

The skeleton as a whole plays a vital role in supporting the animal. From this standpoint, it is possible to consider the skeleton as a problem of structural engineering. Essentially, a land vertebrate is supported on four vertical members, but it is necessary to be a little cautious in making analogies with such man-made objects as bridges. It must not be forgotten that vertebrates are not static, that at times they have to move. Hence, the purely weight-bearing supportive function of the skeleton must of necessity be modified to allow movement of the different parts. With this proviso it remains a useful exercise to analyse the static skeleton as a problem of structural engineering.

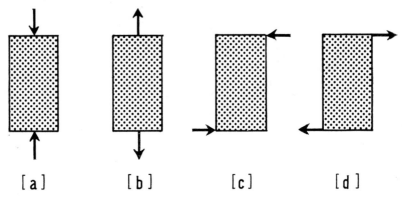

Fig. 19.1. (*a*) Compression, (*b*) tension, (*c*) shear, (*d*) torsion.

As we have seen in Chapter 15, bone is well adapted to withstand compressional and to a lesser degree tensional forces. The failure of bone is hardly ever due to straightforward tension or compression but rather to bending shear or torsion (fig. 19.1). In a limb bone supporting the weight of an animal the load is carried from one bone to another. The main shafts of limb bones are hollow cylinders. For a given amount of material a hollow cylinder will be able to support a greater load in both tension and compression than a solid rod. The compact cortex of vertebrate limb bones gives place to a delicate tracery at the articular ends. The load to be transmitted from one bone to another is distributed

138

along fine struts or trabeculae from the articular surface to the compact cortex of the shaft (fig. 19.2). Because of this tracery the load can be effectively distributed with the limb bones in different relative positions. The struts are aligned along the axes of the forces applied. However, if there were only these trabeculae any compressional forces would produce considerable buckling. In fact G. H. Bell has demonstrated that the delicate trabeculae that run at right angles to the main lines of force act as cross-ties, which not only prevent buckling but also increase the load-bearing capacity of the strut trabeculae (fig. 19.2).

One of the most obvious features of the vertebrates is that they come in many different shapes and sizes. In many ways the size is a major determining factor of their shape. The proportions of the limbs relative to the body depend entirely on the size of the animal provided the habits do not change. If the body is taken as a cube and is increased so that every side is twice the original length, the surface area will be increased four times and the volume eight times. With a threefold increase in the linear dimensions, the area is nine times as great but the volume twenty-seven times. When the hypothetical cubic animal grows to four times its original height, length and width, the surface area increases sixteen times but the volume sixty-four times.

Any increase in size of an animal, if all the parts are equally scaled up, will create a number of serious problems. The cross-sectional area of the bone tissue of the limbs has to carry the weight of the organism. With any increase in size of the animal the cross-sectional areas will have to be increased by the cube. The greater the increase in size the greater the problem of supporting a weight. This means that there is a size of land vertebrate that cannot be exceeded.

There are further problems that arise in any large increase in size. The amount of food required has to be increased by about the cube to maintain the increase in volume, while the occlusal surface of the teeth will be increased by the square. With a large volume the amount of surface area of the animal will be comparatively small and this will effectively prevent excessive heat loss or gain. The diurnal temperature change will have little effect, as the animals concerned would take longer than 24 hours to cool down or heat up a few degrees. For example it has been calculated that the large dinosaurs would have taken 84 hours to alter their internal temperature as much as the diurnal change.

To return to the question of weight bearing, large vertebrates such as elephants, the extinct giant rhinoceros *Baluchitherium* which was 6 m high at the shoulder, and the dinosaurs have massive pillar-like limbs. The limbs were habitually held vertically and not flexed. Recently there has been some controversy regarding such dinosaurs as *Brontosaurus* and *Diplodocus,* and it has been suggested that the cross-section of their bodies indicates that they were habitually land-going and not as generally supposed semi-aquatic living in ancient swamps, the idea being that if

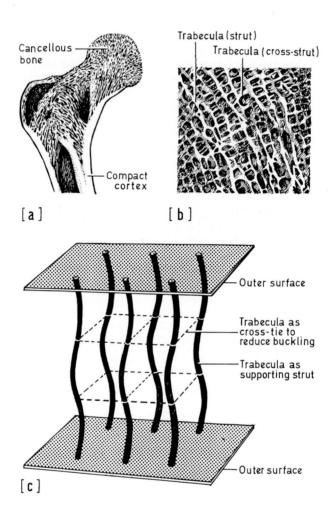

Fig. 19.2. (*a*) Section of human femur to show arrangement of trabeculae distribut-
ing forces from head to compact cortex of shaft, (*b*) detail of trabeculae to
show cross-ties (from L. B. Halstead and J. A. Middleton, 1972. *Bare
Bones—an Exploration in Art and Science*. Oliver & Boyd), (*c*) diagram to
indicate how cross-ties reduce bending or shearing load on trabeculae
(after Bell).

140

they were mainly semi-aquatic like hippos, the body cross-section would have been wider. In fact, the larger the animal the narrower the body so that distribution of the weight to the supporting columns is facilitated. It is easier to support a given weight if the supporting columns are closer together. If their distance apart is too great the strain on the tissues of the ventral surface of the animal would become intolerable.

Although the limbs serve to keep the body off the ground, they are only part of the system of support, and must be considered in conjunction with other parts of the skeleton and body. The limbs articulate with the girdles, which in turn are attached to the vertebral column. The pelvic girdle is fused to the sacral vertebrae by means of a fibrous joint or syndesmosis; the union of the pectoral girdle is by means of a muscular sling (fig. 19.3b). One of the pectoral holding muscles, the serratus anterior, attaches over the dorsal edge of the scapula so that when a pull is exerted by gravity it has the effect of tightening the connection by forcing the scapula against the upper part of the ribs.

The supportive function of the front limbs is not the reason for the muscular attachment of the pectoral girdle to the vertebral column and ribs. When an animal runs or jumps it lands front feet first; the pectoral girdle, therefore, has to be able to take up the shock of impact. A muscular connection will have sufficient 'give' to absorb such sudden forces. At the same time the movement of the ribs that takes place during breathing would be exceedingly difficult if the pectoral girdle was affixed to the ribs by means of a bony union.

The backbone is comparable to the arch of a bridge. The arch of a stone bridge is a compression structure and the individual blocks are wedge-shaped, the central one being designated the keystone. Such a bridge is designed to withstand a load from on top—bridges are meant to carry people and their conveyances. The backbone is similarly arched and is also a compression member. In this case, however, the weight is applied beneath the arch. There is no load applied on the upper surface as is the case in the bridge; furthermore, the individual vertebrae are not wedge-shaped. This would suggest that the neural spines of the vertebrae, the upper part of the compression bar, are likely to be subject to tensional forces. The greater the compression on the vertebral centra or bodies the more likely are the spines to be spread apart. Tendons and ligaments which resist tensional forces run along the dorsal part of the vertebral column and prevent the backbone becoming over-arched. In the dinosaurs the 6th and 7th dorsal vertebrae in the middle of the back are fused and thus strengthen the vertebral column at the point where it is likely to be subject to the maximum of both compression and tension.

The skin, muscles and connective tissue of the ventral surface of the body act as a tensional structure to help maintain the compression member of the vertebral column (fig. 19.3a). The two in conjunction support the viscera, the entire weight of which is transferred to the vertical compression struts of the limbs.

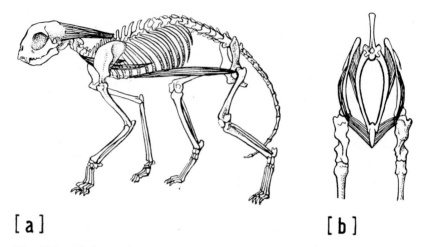

[a]   [b]

Fig. 19.3. (*a*) Cat skeleton to show muscular tensional structures, (*b*) pectoral region of horse to show muscular sling (after Slijper).

The vertebrate body consists of more than legs and guts; there is a head and a tail. The skull is supported by a cantilever. The neck vertebrae act as a simple compression strut, with muscles and ligaments from the skull to the scapula and the neural spines forming the tension member (fig. 19.3*a*). The main tension-resisting structure, especially in animals with large heavy skulls such as horse and cow, is the nuchal ligament, which runs from the cervical neural spines to the skull, and is composed of elastin. This material can be stretched like rubber and becomes so stretched when these animals crop grass and their heads reach the ground. Thereafter, the nuchal ligament springs back and reverts to simply holding up the head. If this task were done by muscles alone much unnecessary expenditure of energy would be required. The tail also is supported on the cantilever principle, with collagenous ligaments and tendons acting to counteract tensional forces. These are only evident when the tail is no longer pendent but scandent.

19.2. *Locomotion on land*

Vertebrates do not merely stand still, they move about and sometimes with an astonishing turn of speed. The skeleton, especially the limbs, is jointed so that there are moving parts. All those concerned with rapid movement have synovial joints, which allow an almost frictionless action. Each joint can be considered in terms of levers. Where the head is nodded at the occipital condyles on the atlas of the vertebral column, the load is moved by a force acting on the opposite side of the fulcrum;

this is a first order lever (fig. 19.4*a*). In a second order lever the load is positioned between the force and the fulcrum; this is the arrangement in the foot, used when one stands up on one's toes (fig. 19.4*b*).

Most movements are the action of third order levers in which the force is applied between the load and the fulcrum (fig. 19.4*c*). The distance of the application of the force from the fulcrum is a measure of the strength of the limb movement. The further the distance the greater the load that can be lifted (fig. 19.4*c*). Limbs of this type are said to exhibit good mechanical advantage and are characteristic of animals with powerful limbs such as ant-eaters and badgers. Fig. 19.4*f–h* shows that if the effort is inclined to the lever arm, so that its *perpendicular* distance from the fulcrum is reduced, the mechanical advantage is reduced, but this

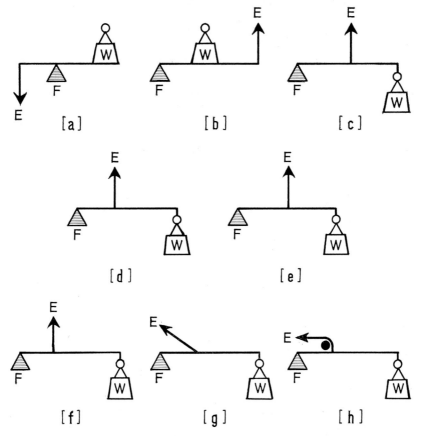

Fig. 19.4. (*a*) First order lever, (*b*) second order lever, (*c*) third order lever, (*d*) low mechanical advantage in order to gain speed, (*e*) high mechanical advantage for strength (*f–h*) increase of mechanical advantage by introduction of a pulley system. E, effort; F, fulcrum; W, load (weight).

can be rectified by employing a pulley system so that the effort is again applied at 90° to the lever arm. The patella for example acts much in this way. Where the force is applied close to the fulcrum only a small weight can be lifted, and this arrangement is described as having a low mechanical advantage (fig. 19.4d). With the force near the fulcrum the arc described by the arm bearing the weight will be much greater for a given shortening of muscle than in a limb showing high mechanical advantage. Limbs exhibiting a low mechanical advantage tend to be light and describe a long arc with considerable rapidity. Animals specialized for speed, such as deer, horses and cheetahs, have this arrangement.

The type of limb can also be described from the point of view of its gear ratio, which is related to the mechanical advantage. It is the distance from the fulcrum or pivot to the point of application of the limb on the ground divided by the distance from the pivot to the point of application of the force, i.e. the insertion of the main limb-moving musculature. The shorter the latter distance is, compared with the length of the limb, the higher the gear ratio. As with motor vehicles once the initial inertia is overcome at the start, speed is achieved by utilizing a high gear.

The speed of an animal is determined by the length of the stride and the rate at which it takes strides. With a high rate of oscillation at the pivot and a longer leg, there will be a higher speed. There are further modifications that increase the length of the stride. In the ungulates the backbone is fairly rigid but the effective stride is increased by a swivelling action of the scapula. Ungulates also have a mechanism for increasing the acceleration. There are suspensory or springing ligaments of elastin running from the posterior part of the metacarpals and metatarsals to the anterior part of the terminal phalanges. When the toes are on the ground, the resulting flexion extends these ligaments, but when the feet are lifted the elastin recovers and gives an extra impetus to the upward movement of the limbs.

In the carnivores the stride is also increased by the swivelling action of the scapula but in contrast to the ungulates the backbone is very flexible. The flexion and extension of the vertebral column of the cheetah adds an extra 10 km h^{-1} to the speed. Ultimately the speed depends on the power output of muscle. This is determined by the surface area of the muscle, which limits the rate of dissipation of heat and the oxygen supply. In fact, the rate of contraction will vary inversely with the linear dimensions, hence larger muscles will contract more slowly than smaller. This means that the larger the body, the slower the rate of stride. Because of this there is hardly any difference in the speed of small animals like hares and large ones such as horses. This situation, however, only applies on level ground.

When an animal travels uphill the limbs have to counteract the additional force of gravity, which will be related to the volume of the

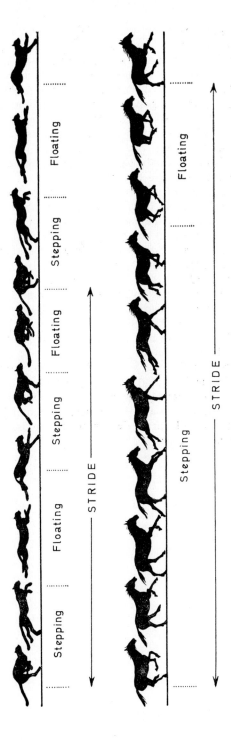

Fig. 19.5. Comparison of cheetah and horse at full gallop, showing difference in proportion of stride in floating and stepping phases (after Hildebrand).

animal. The larger the animal the slower it will be able to move up a steep hill; a dog or cheetah will be able to gallop up a hill while a horse will have to take it at a slow walk. In the same way a heavy lorry will have to slow down and change into low gear on hills while a sports car will be able to race up and over. This same problem of lifting the body against the force of gravity is met in galloping. When animals run at high speed, there are times when all the legs are out of contact with the ground. This is termed the floating phase, in contrast to the stepping phase where at least one leg is in contact with the substrate.

In a gait involving the floating phase there are two competing factors. The longer the floating phase the greater the amount of work necessary to counteract the effect of gravity, but the less often is it necessary for the limbs to accelerate the body. In practice a compromise has to be reached, but it is clear that the larger the animal the smaller the proportion of the floating phase in any stride. With an elephant there is never a floating phase, with a horse at full gallop about a quarter of the stride is floating and with the cheetah a half of it (fig. 19.5). If the horse and cheetah are compared it is seen that the length of their stride is about 7 m. In the horse there is one unsupported period in each stride, whereas in the cheetah there are two. The cheetah achieves speeds of up to 113 km h^{-1}, the horse 68. The enormous speed of the cheetah is due to the timing of the extension and flexion of the backbone in conjunction with the thrusts from the limbs.

In the wild, both herbivores and carnivores can cover huge distances. Caribou herds in the Canadian arctic migrate over 2000 km each year, and packs of African hunting dogs range over 2000 km^2. Such travelling involves not only speed but also endurance. The latter is exemplified by sleigh dogs but it is difficult to determine the endurance of an animal from skeletal remains alone.

flight

20.1. *Movement in air*

20.1.1. *Aerofoils*

Three major groups of vertebrates have succeeded in conquering the air; birds, bats and pterosaurs. The problems of movement in air are the same for all heavier-than-air machines, be they animate or inanimate. There are two problems to be overcome; one is to maintain the object in air and the second is to provide a propulsive force to move it through the air. To understand the basic principles involved, it is easier to consider first the rigid wing of an aeroplane. The cross-section of a wing is curved to give a greater surface area on the upper part. Air flowing over this type of structure, termed an aerofoil, will have further to travel over the upper surface than the lower. This creates an increased pressure below and a decreased pressure above, and the combination of these two pressures results in lift (fig. 20.1*a*). If the front of the aerofoil is raised at an angle to the horizontal, termed the angle of attack, the difference in the distances that air will have to travel along the upper and lower surfaces will be further increased and still greater lift will be generated (fig. 20.1*b*). However, if the angle of attack is increased too much, the smooth or laminar flow over the upper surface separates and turbulence develops, which produces friction causing drag. At this point the wing is no longer able to be supported in air and plummets to the ground; this is termed stalling. There are a number of structures, such as the Handley-Page slot and the alula or bastard wing of birds that help to keep the flow of air free of such turbulent eddies (fig. 20.1*d*).

The leading edge flap is an alternative method of achieving the same results and was developed in both bats and pterosaurs (fig. 20.1*e*). In aeroplanes, bats, and presumably pterosaurs the smooth flow of air was further assisted by a thin turbulent boundary layer generated by small protuberances on the upper surface of the wing, made by the phalanges of the last two groups. The main flow of air is on the turbulent boundary layer and hence there is little friction on the wing itself (fig. 20.1*f*).

The aeroplane wing is rigid and is concerned with lift, but with flying vertebrates, the wing further provides the propulsive force. This is accomplished by altering the angle of the aerofoil and driving it downwards. With a bird moving forwards and the wing being brought down the resultant airstream relative to the wing will be backwards and

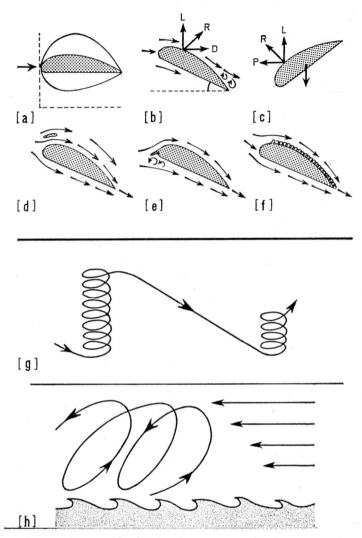

Fig. 20.1. (a) Aerofoil with upper surface convex, air passing round aerofoil travels further over upper surface; this will produce an increased pressure below and a reduced one above, resulting in lift, (b) aerofoil raised at an angle to the horizontal, the angle of attack; this will increase the pressure difference between the upper and lower surfaces, thus increasing lift (L). The friction of the air will create drag (D); the resultant force (R) is calculated from the lift and drag, (c) aerofoil is driven downwards at an angle, and the forces on the aerofoil are utilized to produce propulsion (P) as well as lift, (d) laminar flow over the aerofoil is facilitated by presence of an anteriorly placed slot, (e) leading edge flap produces an effect similar to a slot, (f) a small protuberance on upper surface generates a thin turbulent layer, which acts as a kind of lubricant helping to maintain overall laminar flow. Soaring: (g) static soaring; rising in thermals, gliding down to rise again in a further thermal, (h) dynamic soaring; rising by exploiting the gradient from wind shear, thereafter gliding downwind to turn and soar upwards on the gradient.

upwards. The forces on the wing are thus translated into forward propulsion and lift, instead of lift and drag (fig. 20.1c).

20.1.2. Gliding and soaring

Flying vertebrates spend a large proportion of their time in the air gliding. The wing loading, i.e. the animal's weight (measured of course in force units) over the wing area, determines the sink speed. Hence, the larger the area of the wing in relation to the weight of the body the lower the sink speed. This means that with a lower wing loading, the animal will be able to remain airborne at lower speeds. The albatross has a wing loading (in *force* units per unit area) of 160 N m^{-2}*, man-made gliders of 200 N m^{-2}; for the giant pterosaur *Pteranodon*, with a wingspan of 8·2 m, it has been calculated by Bramwell as 36 N m^{-2}. The calculation suggests that *Pteranodon* was an exceedingly slow glider with its optimum performance in the range of 5·7 to 7·7 m s^{-1}. When flying at 6·7 m s^{-1} *Pteranodon* would have sunk at the rate of 0·5 m s^{-1}. From this study it is evident that *Pteranodon* was one of the most efficient gliding machines ever constructed. Indeed, before the advent of helicopters, a pterodactyl aeroplane was built with advice from the palaeontologist D. M. S. Watson, and was tried out by the British army during the 1920's. It had the astonishingly low stalling speed of 40 km h^{-1}.

The outline of the wing which gives the area is described in terms of aspect ratio, determined simply by dividing wing length by wing breadth. A wing of high aspect ratio is long and narrow as in the albatross (20) and swifts, one of low aspect is short and broad as in the vulture (7). The former stall at high speeds, the latter at lower speeds—a useful attribute in an organism that exploits rising air masses in thermals. Slow-flying birds, with short broad wings, splay out the primary feathers so that the wing as a whole is slotted. This allows eddies at the wing tips to slip through and thus prevents stalling due to increased drag.

In a similar way long distance flying bats have wings with high aspect ratios, short distance ones with low. The different types of wings among the pterosaurs must also indicate similar differences of habit.

These considerations of gliding have been in relation to still air. The aerodynamic qualities of the animal and its wings will determine the rate of sink. If the air rises at a greater speed than this, the animal will be carried upwards; it will soar. Rising currents of air occur in two situations. First, as a current of air meets an obstacle it will rise to pass over it, so that where the ground slopes, seen at its extreme by sea cliffs, there is an upward movement of air, which can be exploited by any gliding object. Utilization of such currents is known as slope soaring.

---

* A force of 10 N supports a mass of 1 kg, so these figures can be interpreted, in terms of mass supported, as 16 kg/m^2 . . . and so on.

Only the windward side of slopes can be used; this type of soaring is seen to best advantage in gulls and especially fulmars on the coast.

Secondly, over the land where warm bubbles of air or thermals rise, birds and man-made gliders use these currents to obtain lift. In India and Africa, it is a common sight to see vultures slowly wheeling high in the sky or thermalling. Pennicuik records how vultures travel over a hundred kilometres by cross-country soaring, rising 500–1000 m in a thermal and then gliding 6–12 km to rise again in another thermal (fig. 20.1g). In fact, thermals are not randomly distributed but frequently occur in lines. By using such 'thermal streets', vultures can prolong straight glides of up to 30 km without any appreciable loss of height.

Slope soaring and thermalling are described as static soaring in contrast to dynamic soaring, which exploits the gradient in air speed over a surface. As wind passes over a surface, such as the sea, friction will slow it down so that there will be a speed gradient from the interface, where the wind shear will be at its maximum. This will gradually slacken until at about 50 m the full air speed will be attained. The rate of change of wind speed is greatest near the surface. The bird by facing into the wind is able to exploit this gradient and as it rises it gains speed. When the gradient is too weak, the bird will turn and glide downwind. At sea level the bird is able to slope soar using the windward side of the waves, and then it turns upwind to take advantage of the gradient once again (fig. 20.1h). The albatross, and presumably the ocean-going pterosaurs, soar by a combination of static (slope) and dynamic soaring.

A further aspect of gliding and soaring is that of manoeuvrability. Flying animals twist and turn and when an animal turns a greatly increased strain is put on the skeleton. For example, the pigeon, with a wing span of 0·67 m, has a minimum turning circle with a radius of 3·4 m, and experiences a centripetal force of about 40 newton per kilogram ($=4\,g$), much the same as for man-made gliders. Bramwell has calculated that for the pterosaur *Pteranodon*, with a wing span of 8·2 m, the turning circle had a radius of only 5·2 m. This means that *Pteranodon* was capable of utilizing the faster central portions of thermals instead of the outer rings used by vultures and man-made gliders, although it is not known whether it actually ever did so in life. The centripetal force exerted on the skeleton of *Pteranodon* would have been a mere 14·7 newton per kilogram (1·47 g), which is just as well since its bones were paper-thin. A curious feature revealed by Bramwell's computer analysis of the flight characteristics of *Pteranodon* is that whatever the rate of climb the angle of bank would have been 23°.

20.1.3. *Powered flight*

Gliding and soaring depend on the opportunistic exploitation of the natural movement of air. Birds and bats also propel themselves through the air by their own muscular power as in the past did pterosaurs. In this case the wings have to provide not only lift but also the propulsive force to drive or propel the animal through air.

In birds there are two types of powered flight, one that can only be maintained for a matter of seconds and is concerned primarily with take-off, and the other concerned with sustained flight. The wing has

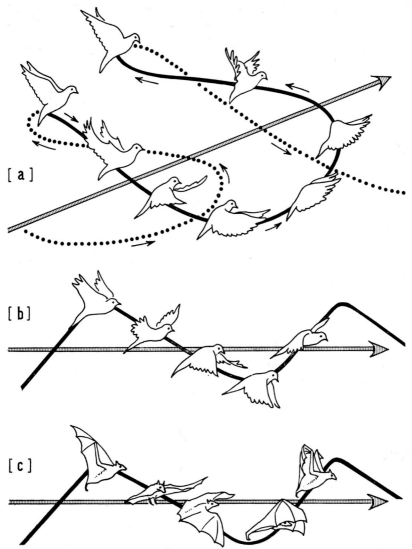

Fig. 20.2. Powered flight. (a) Bird take-off action, the bold figure of eight indicates the movement of the wing tips, the shaded straight line the actual flight path, (b) sustained flight in birds, (c) sustained flight in fruit bats.

to perform two separate functions, lift and propulsion. The inner part of the wing is concerned with lift while the outer part acts as a kind of propeller. The movement at take-off involves the outer part of the wing describing a figure of eight (fig. 20.2a). From a position with the wings raised, they are brought both downwards and forwards thus generating lift. On the upstroke the wing is adducted, folded and flexed thus reducing the area that is driven against the air, then there is a flick both upwards and backwards by a forward and upward movement of the humerus accompanied by an extension of the wing. This last movement produces a powerful forward thrust. This type of flapping flight is exceedingly exhausting and a bird can only maintain this action for a few seconds.

Once the bird is airborne and during sustained flapping flight the pattern of movement is somewhat modified. The inner part of the wing is still concerned primarily with lift and the outer with propulsion. The upstroke is much simpler, with only a minimal backwards flick of the primary feathers. Furthermore the upstroke is a fairly passive action and the feathers part to allow the air to pass freely through. Both lift and propulsion are generated during the downstroke, which does not have a marked forward component as in the action at take off (fig. 20.2b).

The pattern of movement in the sustained flight of bats is comparable to that seen in birds (fig. 20.2c). The main lift and propulsion are produced during the downstroke. The wing membrane is held taut by the four fingers. During the upstroke the wing is strongly flexed and folded so that its resistance to the air is reduced to a minimum. The take-off in bats is by letting go from the perch and launching into the air, although since they hang by their feet, they must turn through at least a right angle. When fruit bats, with high aspect ratio wings, are grounded they can become airborne but are incapable of rapidly gaining height. For many tens of metres they will only gain a few metres. This contrasts with the insect-eating bats which are capable of rapid twists and turns as well as sudden upward swoops.

20.2. *Bird skeleton*

The conquest of the air by birds has resulted in numerous modifications of the skeleton. The success of the birds over bats and pterosaurs is due partly to the possession of feathers, which have the advantage over a membrane of being able to regenerate and to part when in contact with any sharp obstruction. Perhaps one of the most critical structural modifications that makes birds such effective fliers is that when they are grounded the organ of flight can be folded out of the way against the body. At the same time, birds are efficient cursorial bipeds in marked contrast to bats, which are exceedingly vulnerable once grounded, as must also have been the case with the pterosaurs.

The skeleton of the bird wing is characterized by the reduction in the distal elements. The first digit is retained and supports the bastard wing or alula. This, when opened away from the wing, makes a Handley-Page slot. The main flight muscle, the pectoralis, originates on the deep keel of the sternum. Since this muscle not only lifts the animal but propels it through the air, the entire pectoral region requires appropriate strengthening. The coracoid provides a compression member between the humerus and sternum, and the fused clavicles (furcula, merrythought or wishbone) further assist in the bracing. The anterior thoracic vertebrae are fused. Additional attachment for the holding musculature from the scapula is provided by the posteriorly directed uncinate processes on the ribs (fig. 20.3).

The pelvic girdle is also highly specialized, with the girdle fused to the vertebral column to form the synsacrum. This strengthening is to enable the animal to withstand the shock of landing as well as the force of take-off. The vertebral column ends in the ploughshare bone or

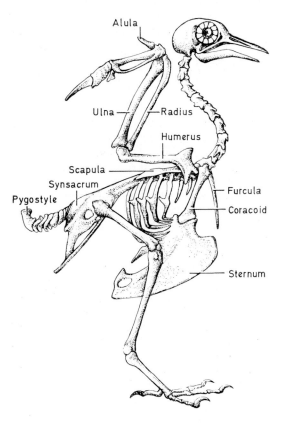

Fig. 20.3. Bird skeleton.

pygostyle, which is concerned with the support of the tail feathers which play an important role in braking, hovering and other aerial manoeuvres.

The most notable feature of the bone of birds is that extensions of the lungs penetrate their interior; this type of bone is termed pneumatic. Such bone is known in pterosaurs and when their remains were first found in the eighteenth century they were identified as the bones of geese. These light bones, however, contain internal cross-wise supporting struts, which are paralleled in aeroplane wings.

The earliest known bird, *Archaeopteryx,* from 140 million years ago, did not possess pneumatic bone and, in contrast to the later birds, had a long bony tail and teeth. The tail and its feathers must have acted as an automatic stabilizer, which would have corrected any undue tendency to pitch. *Archaeopteryx* was probably more of a parachutist than a flier, but it must have been stable. However, its stalling speed must have been unduly high. With the subsequent loss of the tail, the animal would have lost weight and at the same time have become unstable. This would have allowed a lower stalling speed, a prerequisite for manoeuvring. Such changes, which also occurred during the evolutionary history of the pterosaurs, were only possible with concomitant improvements in the brain, in the cerebellum, which is concerned with muscular co-ordination and balance.

20.3. *Bat skeleton*

Exactly the same problems regarding the strain on the pectoral region are present in the bat, but have been overcome rather differently. The scapula is joined to the manubrium of the sternum by the strong rod-like clavicle. The holding sling musculature, the serratus anterior, is greatly developed. Again the main flight muscle is the pectoralis, which originates not on a bony keel but on a ligamentous sheet of connective tissue. The full extension of the arm at the elbow is prevented by the ulnar sesamoid bone and also by the leading edge flap which runs from digit II to the shoulder. Most of the pollex is free and acts as a grappling hook, but a short proximal phalanx gives some support to the leading edge (fig. 20.4). The patagium or wing membrane is supported by all five digits. A complex system of tendons ensures the rigidity of the patagium during the downstroke. The membrane has fibres of elastin which always ensure a smooth surface and outline when the wing is flexed on the upstroke or at rest. Collagen fibres resist tensional forces, especially during the downstroke. Finally there are isolated strips of patagial muscles which contract to provide tensional forces in an anteroposterior direction. In marked contrast to birds, the bats have the ability to alter the camber of the wing, which gives them considerable aerodynamic versatility.

The pelvic region is greatly reduced and the hind limb is rotated, with the patella posteriorly placed, as an aid to the hanging habit of bats (fig. 20.4). The main function of the hind limbs is to support the

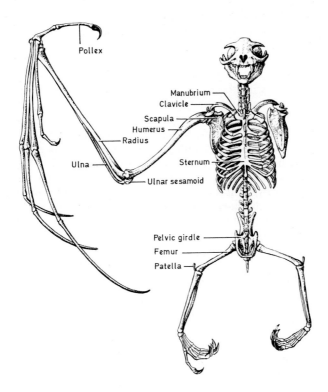

Pollex

Manubrium
Clavicle
Scapula
Humerus
Radius
Ulna
Ulnar sesamoid
Sternum

Pelvic girdle
Femur
Patella

Fig. 20.4.   Fruit bat skeleton.

weight of the animal as it hangs. The other important role of the hind limbs is to control the camber of the wing membrane during flight. In fact, during normal sustained flight, the rhythmic action of the hind limbs is an integral part. Finally the bones of bats are not pneumatic. As with most mammals, bats possess a furry covering which acts as an insulator and helps in the maintenance of a constant body temperature, at least during their periods of activity.

20.4. *Pterosaur skeleton*

Pterosaurs are popularly described as flying reptiles. A. G. Sharov has recently described one which he named *Sordes pilosus* ('hairy filth'); the preservation of this animal is remarkable in that the furry or downy covering can be clearly discerned on the body and hind limbs. It therefore seems that unlike most reptiles, which are scaly and cold-blooded, some at least of the pterosaurs had a covering that must have acted as an insulator and, hence, it is inferred that they were warm-blooded.

155

The wing was formed by a membrane attached to a greatly elongated fourth finger and held at the ankle. In the primitive pterosaurs there was a long stiff tail, which acted as an automatic stabilizer, and there was a further membrane between the legs attached to the tail. The probable ancestor of the pterosaurs, the small lizard-like *Podopteryx*, was a parachutist with the main gliding membrane being this posterior one with only a small membrane between the knee and elbow. The bones of the pterosaurs were pneumatic. The advanced pterosaurs lost the tail and also the teeth, which were heavy dense objects.

All pterosaurs showed a number of specializations in the skeleton, the most striking of which were in the pectoral region. The pectoral vertebrae were fused into a solid block, the notarium, and the scapula was not attached by a muscular sling, but directly to the notarium by a fibrous joint or syndesmosis. The coracoid was similarly attached to the sternum, so that the pectoral arch formed an efficient bracing mechanism (fig. 20.5). There was not a marked keel to the sternum, which was probably similar to that of the bat in that the pectoralis originated on a ligamentous sheet.

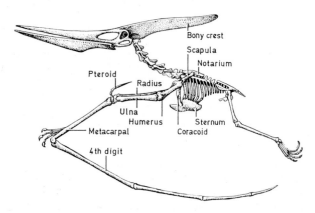

Fig. 20.5   Skeleton of pterosaur *Pteranodon* (after Kuhn).

The wing bones comprised the humerus, which from an examination of the articular surface of its head is seen to have been held in the glenoid cavity at an angle of 20° to the vertical and 19° posteriorly. From the distal part of the humerus to the proximal end of the first phalanx, there was a tendon running along the leading edge of the radius and the proximal part of the metacarpal to insert on the first phalanx on its dorsal surface. The effect of this tendon was to counteract tensional forces as well as a possible tendency of the outer part of the wing to twist. The leading edge of the wing forms an upward curve, which further acts to reduce the tendency to twist. At the wrist joint there was a small spike of bone which had a ball and socket joint. This pteroid bone supported a small membrane extending from the wrist to the base of the neck. Movement of the pteroid bone changed the position of the pteroid membrane, which acted as a leading edge flap, and was thus able to alter the camber of the wing.

The joint which allowed the greatest degree of freedom of movement was the metacarpal–first phalanx. At this point the outer part of the wing could be folded back on itself. This is the point where changes in the configuration of the wing were determined. When the animal was grounded the wing was folded back across the animal's back at this joint. It was at this same place that the clawed fingers survived. These were presumably of use in progression on land, as in fruit bats.

A further feature of the skeleton, which characterized *Pteranodon*, was the enormous bony crest, extending from the back of the skull (fig. 20.5). The formation of this extra bony growth would seem at first sight to be a retrograde step in the evolutionary trend to reduce weight. In fact, it was a thin blade only some 3 mm thick. The function of this blade has been worked out by Bramwell and Whitfield from wind tunnel experiments. Its effect would be to balance loads on the beak as the animal turned its head during flight. This is very important when searching for fish. The weight of this blade would have been much less than the amount of necessary neck musculature. A turn of the head of 70° has little effect on the couple of the head on the neck with or without the crest. When the head was turned through 90° the crest did in fact act as a balance to the beak. At such turns the neck muscles would have been fully extended and at such a critical juncture the crest would have fulfilled an important role.

Many aspects of the flight of *Pteranodon* are now reasonably well understood but in order to feed the animal had both to catch and to carry its food, and this would certainly have added to its problem of weight. The centre of gravity and the centre of lift of the gliding *Pteranodon* coincide. There is a gap in the mandible, which has been interpreted to mean that *Pteranodon* possessed a throat sac. Furthermore, the jaw articulation has a spiral arrangement which allows the two sides to move apart. This arrangement is identical to that in the pelican, which has a considerable throat-sac, and by the same token it is inferred that *Pteranodon* had a comparable structure.

It is necessary to consider what happened when a *Pteranodon* caught a fish. If an individual weighing 16·6 kg caught 4 kg of fish, both the centre of gravity and the stalling speed would be affected. If the fish were held in the feet the centre of gravity would be moved back some 330 mm, at which point the animal would cease to be airborne. If the fish were held in the throat sac with the head pointing forwards the centre of gravity would be moved forwards 70 mm, which is tolerable; with the head raised the movement would be 25 mm. This simply means that

Pteranodon both caught and carried its fish in its beak and did not fish in the same way as the fish-eagle or the fishing bat.

From a detailed study of the skeleton alone, for that is all that there is, a great deal regarding the behaviour of an extinct flying vertebrate has been determined.

trauma and disease

21.1. Fracture healing

The skeleton has the remarkable property of being able to repair itself completely, by regenerating dead and broken parts. The repair of a fracture of an adult human limb bone is such that after a year there may be no sign of any break having occurred. A fracture in the arm will take only three months to heal, one in the leg six.

When a bone is broken, the surrounding soft tissue is similarly ruptured. This is especially the case with blood vessels, which results in blood flowing into the damaged area with the subsequent formation of a clot or haematoma callus. The consequence of the damaged region being sealed off by the clot is that the blood supply to the bone adjacent to the fracture is cut, so that the osteocytes die and the bone becomes necrotic (fig. 21.1a). The process of repair begins with the formation of a callus, the tissue of which grows around and encloses the damaged part. This develops from two sources, cells from within the marrow and cells from the periosteum. Osteogenic cells from the marrow invade the blood clot and build bony trabeculae which eventually unite the broken ends of the bone. This development is known as the internal callus, in contrast to that formed from cells of the periosteum, the external or bony callus.

Cells from the periosteum proliferate and form a collar of tissue around each of the broken ends of the bone. These collars grow and meet each other to enclose the fracture completely. The outer surface comprises fibrous tissue but the main part of the callus is made up of cartilage, and this is the fibrocartilage callus. Capillaries invade the callus and at the margins, osteoblasts form new bone on the surface of the dead necrotic bone (fig. 21.1b). The cells that differentiate in a non-vascular environment become chondroblasts and chondrocytes. Cartilage thus forms in the parts of the callus furthest from the fracture surface. This cartilage is gradually replaced by bone in the same way as in normal endochondral ossification of bone.

Bone is also formed beneath the extended periosteum so that the two ends of broken bone are united by a thin bony shell (fig. 21.1c). Replacement of the cartilage proceeds from the edges of the callus and also by trabeculae spreading inwards from the outer fibrous layer of the callus. As this process continues the trabeculae of bone in the internal callus effect a further union of the broken bones. The fracture at this stage

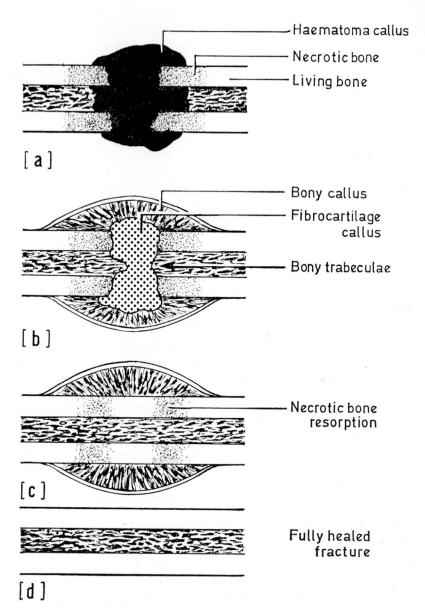

Fig. 21.1. Diagram of major stages in healing of bone fracture. (*a*) First stage in process with formation of blood clot or haematoma callus and death of bone adjacent to the fracture, (*b*) replacement of clot by fibrocartilage callus and growth of bony collars from periosteum, which unite (as in this diagram), (*c*) bony trabeculae join up within main cavity of shaft and replace the cartilage, the trabeculae of the bony callus also replace the cartilage but, in the region of the cortex, new compact bone is formed, (*d*) in the final stage the area of necrotic bone is remodelled, so that it is replaced by living bone.

is united by new bone some of which is attached to necrotic bone. Between the trabeculae affixed to dead bone, resorption of the necrotic bone takes place and as it is removed new living bone is laid down to replace it. Eventually all the dead bone is given over to the living.

In the final stage in the repair process the opposite ends of the formerly broken bones develop compact bone which thus re-establishes the cortex (fig. 21.1*d*). The surrounding trabeculae of the bony callus are completely resorbed, so that the final result is indistinguishable from the limb bone before its traumatic damage.

This type of perfect repair can only occur if the limb concerned is immobilized and the bones are correctly set. Without this the muscles and tendons will pull up the distal fragment so that at the end of the formation of new bone the limb may be significantly shortened.

21.2. *Skeletal malformations*

The normal skeleton can be modified in other ways than by trauma. Perhaps the most dramatic malformations are the result of developmental disorders producing teratoma or monsters. The skull illustrated in fig. 21.2 belonged to an Indian boy killed by a snake bite when he was 4 years old; he had a normal body, with the addition of a further skull fused upside down on the top of his head. This is likely to have

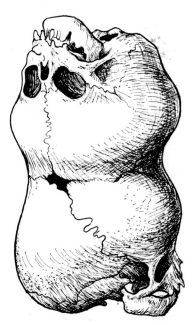

Fig. 21.2. Craniopagous skull of four-year-old male child (Hunterian Museum, Royal College of Surgeons, London).

161

been the partial development of a Siamese twin, although the rest of the body was not developed.

Generally speaking such foetuses do not survive birth. This is so with, for example, cyclops babies; a specimen of which is illustrated in fig. 21.3. The cyclops condition is found in horses, sheep, and other species, besides man. There is a single centrally placed orbit, although the eye may have two pupils. Similarly other organs that are normally paired

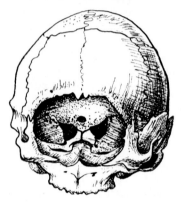

Fig. 21.3. Skull of newborn human cyclops (Odontological Museum, Royal College of Surgeons, London).

are fused. This is an extreme example of developmental disorder. The presence of supernumerary limbs is less serious as these can be removed at birth. There is little that can be done with a situation like that which arose with the foetuses of women who had been taking the drug thalidomide, which resulted in the enormous reduction in length of the proximal portion of the limbs. Perhaps the most extreme of all skeletal developments is found in ovarian cysts. Some 90 per cent of these growths contain perfectly formed isolated membrane bones and teeth. These are believed to be a kind of virgin birth in which the foetus is quite disorganized. The objects formed in these cysts seem to be associated with dermal structures. A similar development occurs in the teeth which very occasionally form in the testes of horses. All these gross malformations can be thought of as rare accidents.

Genetic malformations also occur, such as achondroplasia in which the head and trunk are normal but where the limbs, especially the legs, are foreshortened. Circus dwarfs are generally achondroplasics. One of the most lethal mutants is that of the grey lethal in mice, in which there is no osteoclast activity so that no bone remodelling can occur. The teeth do not erupt because the bone enclosing them in their bony crypts cannot be resorbed.

Further major changes in the skeleton are effected by excessive production of the growth hormone somatotropin, which leads to the condition of acromegaly. Growth continues after the epiphyses have fused to the diaphyses, and so produces the characteristic massive mandibles and huge hands and feet. Virtually all the so-called giants suffer from this over-activity of the pituitary. With insufficient growth hormone there is a retardation of the normal growth processes so that a pituitary dwarf develops. These people are normally proportioned but on a reduced scale. The other type of dwarf, the thyroid dwarf or cretin, is produced by insufficient thyroid hormone. Unless treated the child retains infantile proportions throughout life and, as the popular name implies, is mentally retarded.

Bony growths on the skeleton, which are commonly seen as convex lumps on the skull, are benign growths or osteomas. There are cases of primary sarcomas or malignant cancers of the bone. Frequently when cancers develop in the soft tissues, only much later do they come to be lodged in the hard tissues.

A once common skeletal deformity was rickets. This is a deficiency disease resulting from a lack of vitamin D. This causes a lowering of the inorganic phosphate in the plasma with a subsequent failure of calcification in the developing skeleton. The calcium and alkaline phosphatase levels remain normal however. There is no calcification in the epiphyseal plate, so that the cartilage continues to proliferate without being replaced by bone. The bony trabeculae become covered with uncalcified matrix. The effect of this is that the limb bones are unable to support the weight of the child and so they buckle. The femora curve forwards and the tibiae anteriorly and laterally thus distorting the entire posture. The adult will retain such a posture for the rest of his or her days. Rickets is a disease found only in children; the comparable ailment in the adult is known as osteomalacia, in which the posture is not altered but the matrix of the bone becomes progressively de-calcified. This disease was well known in Imperial China where it was a result of a diet of low calories, low protein and insufficient calcium and phosphate intake. It is not uncommon today in the Indian subcontinent.

At the opposite end of the spectrum is the condition known as marble bone or osteopetrosis. There is a great increase of hard tissue at the expense of the soft, although there is no change in the outward appearance of the bone itself. There is some evidence that this trait is inherited. The calcium and phosphate levels in the blood remain normal, and the main effect is that the bone is remarkably brittle. It is noteworthy that the manatees and dugong both have dense massive bone not unlike this, although in this case it is certainly not pathological. With these aquatic herbivores the condition is best described as pachyostosis.

21.3. Skeletal degeneration
21.3.1. Osteoporosis

When a limb is immobilized, or if a person is subject to an extended period of weightlessness, the amount of bone becomes significantly

163

reduced. The outer dimensions of the individual bones do not alter but the actual amount of bone tissue is reduced. The compact cortical bone is thinned and so are the trabeculae of the cancellous bone. In some instances the actual mass of bone may be reduced by as much as 75 per cent without any change in the external dimensions. The consequent increase in the porosity of the bone is described as osteoporosis.

Such examples of atrophy caused by disuse occasion little controversy but osteoporosis is a disease that is not well understood and arguments regarding its nature still continue. Some authors contend that it is not a real disease but is simply part of the normal ageing process. Certainly it is commonly found among elderly persons, and there is a significant increase of osteoporosis in women between 40 and 50 years of age. This is often spoken of as post-menopausal osteoporosis, and is attributed to the cessation of oestrogens. There is an increase in bone resorption, with large amounts of urinary calcium being recorded. Administration of oestrogens will reduce the urinary calcium. This is not by any means the whole story. The bones of old men also become osteoporotic and it cannot be simply a matter of oestrogen especially in view of the observation that negro women are not as prone to osteoporosis as white women. The general consensus is that osteoporosis can be a pathological condition but is also part of the normal ageing process. There is some evidence that changes occur in the collagenous matrix and that these contribute to the onset of osteoporosis among the aged.

21.3.2. Osteoarthritis

Osteoarthritis is one of the most familiar gradual accumulative diseases of the skeleton associated with ageing. It has been suffered by the vertebrates for at least 200 million years. In spite of its enormous antiquity it is still not well understood and a cure still eludes the medical profession. Osteoarthritis is a disease of the joints, and one of the first stages of its progression seems to be an alteration in the nature of the synovial fluid occasioned by disruption of the blood supply. When an arthritic joint is opened the joint space is full of copious amounts of synovial fluid and blood. This fluid is no longer an effective lubricant and the articular cartilage becomes plucked off. Collagen fibres project into the joint space rather like the pile of a carpet. Eventually all the underlying hyaline cartilage is removed by the friction in the joint. Thereafter, bone articulates directly on bone—a most painful action. The grinding of bone on bone produces a highly polished surface rather like ivory, and this is termed eburnation.

There is a reaction to this abrasion and around the margins of the joint and in the attachment of tendons and ligaments there is a reactive proliferation of bone. This lipping around the margins of the joint is

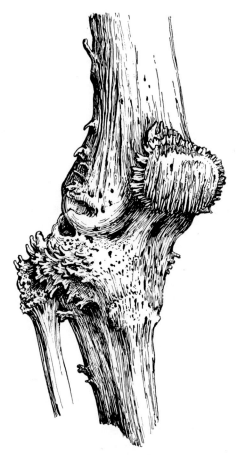

Fig. 21.4. Anchylosis of human knee (from L. B. Halstead and J. A. Middleton, 1972. *Bare Bones—an Exploration in Art and Science*. Oliver & Boyd).

known as osteophytosis and is indicative of an advanced stage of osteoarthritis. Unless the joint is kept moving, the reactive bone growth around the margins will meet up and complete fusion or ankylosis will take place. Once this has happened there can be no subsequent movement, but at least at this stage there is unlikely to be much pain either. A completely fused knee joint from an amputated limb is illustrated to show the final stage of osteoarthritis (fig. 21.4).

In advanced cases of osteoarthritis, surgical treatment is now possible and artificial bearing surfaces of plastic or steel can replace the defective joints (arthroplasty), or artificial fusion (arthrodeses) can be undertaken.

Although arthritis is associated with ageing, its localized occurrence in the body is often a measure of a person's occupation. Patagonians hunting with bolas (two stones joined by a cord) developed arthritis in the shoulder, the ancient Minoan bull-leapers showed osteophytosis of the vertebrae. Arthritis of the fifth lumbar vertebra is common among Europeans, and is attributed to the perpendicular thrust occasioned by sitting in cars and spending long hours slumped over desks and viewing television. Be that as it may, the Anglo-Saxons suffered from the identical ailment without cars or television. The giant short-necked barrel-shaped plesiosaurs developed arthritis in the neck when they reached old age and so did the cave-bears living in Europe during the Ice Ages. The first complete skeleton of Neanderthal man to be described was of an old individual, and his skeleton was severely distorted by osteoarthritis of the vertebrae. This disease-ridden skeleton led to all restorations of Neanderthal man being given a posture which no healthy individual would ever have supported. This long-standing misinterpretation of Neanderthal man was a consequence of palaeontologists not recognizing pathological conditions.

21.4. *Infections of the skeleton*

Various infections in the body can affect skeletal tissues. When sores on, say, the shins spread, and the periosteum is infected the underlying bone will be similarly affected. The periostitis that follows produces a characteristic pattern of necrotic bone on the surface. Should the infection penetrate the bone so that the marrow is invaded the condition is known as osteomyelitis.

There are a number of diseases which attack the skeleton. Leprosy does so in a very obvious manner, although it begins as a general disease of the tissues, beginning with a loss of sensation at the extremities followed by tissue loss. It is a saddening sight to meet lepers with the digits having rotted away to the merest stumps—the palm held out for alms. Another dramatic disease that destroys the skeleton, or rather selected parts thereof, is syphilis. Only in its very last stages does this most drastic of the venereal diseases attack bone. The nose rots away and the front incisor teeth drop out. As abscesses develop, these invade the skull bones which become eroded away with fistulous openings for pus to escape and large areas of bone becoming necrotic. There is a certain amount of new bone growth which gives a knobby surface to the bone. This terminal condition of syphilis is very rarely encountered today, but in the past it was the fate of many low and high ranking people in Britain, before penicillin or heavy metal therapy.

A disease about which very little is known and which is fortunately extremely rare is leontiasis. In this condition there is a gradual proliferation of the bones of the face. The nose is obliterated and the eyes are pushed out of their sockets leading to blindness. Leontiasis causes little discomfort until the terminal stages but is astonishingly disfiguring to the unfortunate sufferers, as can be attested by photographs of victims.

Besides bone, the teeth are also subject to disease. Mainly as a consequence of too much carbohydrate, and especially sugar, in the diet teeth are susceptible to decay or caries, which, on occasion, can give intense pain. Furthermore infection can spread down the periodontal membrane —the Sharpey's fibres of the syndesmosial joint of the tooth in its socket. An abscess can then develop at the apex (root) of the tooth which will lead to bone resorption to accommodate it. One of the worst examples of dental abscesses and caries is found in the 40 000 year old fossil skull of Rhodesian man. This emphasizes that such diseases have a long history and are not confined to advanced societies.

21.5. Pathological calcification

Every organ of the body can be calcified. This most readily occurs following trauma to the affected part. Muscle frequently ossifies some three weeks following contusion. There is a rare disease, myositis ossificans, in which ossification of muscles progresses throughout life, involving the muscles of the neck, shoulders, back, hips and thighs. One of the most extreme examples of this is a skeleton in the Hunterian Museum of the Royal College of Surgeons in which this disease had gone so far that hardly any joint was capable of movement (fig. 21.5).

Implants of material can also induce bone formation or calcification. A cautionary tale is provided by the synthetic sponge that was used in Czechoslovakia for breast prostheses, which was later discovered to induce bone formation. H. Selye has described innumerable experiments in which he managed to induce calcification throughout the skins of rats and selected organs such as the eyes. In fact he seems to have travelled through the whole of the body calcifying the individual organs. For every deliberately induced calcification he has pointed to the same situation known in humans as a pathological condition.

21.6. Calculi

Mineral deposits can form discrete bodies in parts of an organism with the space to accommodate them. The most familiar mineral deposit is calculus which is removed when the dentist scales the teeth. It is made up of calcium phosphate and other minerals and encloses bacteria. It is concentrated on the lingual surface of the lower incisors where the salivary glands open into the oral cavity.

With over-active parathyroid, the plasma calcium level is raised and the large amount of calcium being excreted frequently leads to the formation of kidney stones. These are composed of uric acid, calcium and magnesium salts and urates. Gallstones are made up of cholesterol, calcium salts and bile pigment.

Finally, objects can be impregnated with calcium salts or enclosed in mineral deposits to seal them off from the rest of the body. Examples

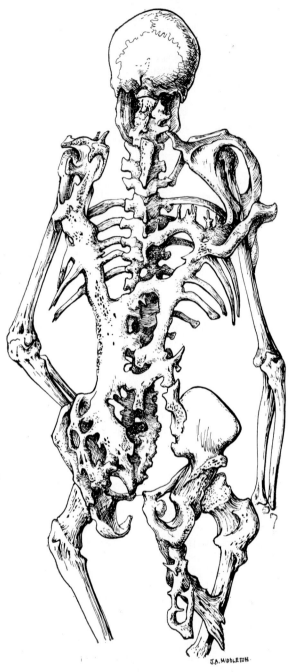

Fig. 21.5. Skeleton of 39 year old man showing advanced condition of myositis ossificans (Hunterian Museum, Royal College of Surgeons, London).

of this include the rare examples of foetuses which die in the womb and are not aborted. These embryos are retained in the mother and are impregnated with calcium salts to form a stone child or lithopedion, looking rather like a piece of primitive sculpture. Calculi will form around foreign objects introduced into organs, the most bizarre of which include tie-pins, peas and chewing gum introduced into the bladders of young men. Other foreign objects such as the cysticercus (bladder worm) stages of tapeworms, embedded in muscle, will also become calcified in man after five or six years.

APPENDIX I
further reading

Since 1967 there has been an international journal devoted entirely to hard tissues—*Calcified Tissue Research*. From 1963 there has been an annual European Symposium on Calcified Tissues, the first organized by an *ad hoc* committee of the 'Bone and Tooth Society'. The proceedings of these international meetings have been published as separate books or as Supplements to *Calcified Tissue Research*. The work reported in the journal and these Symposia is of a highly technical nature and in large measure represents the major source material for this book.

To a large extent books which deal with molecular and cellular aspects also cover the tissues. Hence, to avoid undue repetition I have listed the further reading for Parts I and II together.

PART I: MOLECULAR AND CELLULAR ASPECTS
PART II: THE TISSUES

BOURNE, G. H., ed. (1972), *The biochemistry and physiology of bone*, Vol. 1 (Academic Press).
CLOWES, R. (1967), *The structure of life* (Penguin Books).
GAUNT, W. A., OSBORN, J. W., and TEN CATE, A. R. (1971), *Advanced dental histology* (John Wright).
HALSTEAD, L. B. (1969), *The pattern of vertebrate evolution* (Oliver and Boyd).
HAM, A. W. (1969), *Histology* (Lippincott; Blackwell).
HANCOX, N. M. (1972), *Biology of bone* (Cambridge University Press).
MCLEAN, F. C., and URIST, M. R. (1968), *Bone* (University of Chicago Press).
MILES, A. E. W., ed. (1967), *The structural and chemical organisation of teeth* (Academic Press).
SCIENTIFIC AMERICAN (1965), *The living cell* (Freeman).
SIMKISS, K .(1967), *Calcium in reproductive physiology* (Chapman & Hall).
SLAVKIN, H. C., ed. (1972), *The comparative molecular biology of extracellular matrices* (Academic Press).
VAUGHAN, J. (1970), *The physiology of bone* (Oxford University Press).

PART III: THE SKELETAL SYSTEM

ALEXANDER, R. M. (1968), *Animal mechanics* (Sidgwick and Jackson).
ALEXANDER, R. M. (1971), *Size and shape* (Edward Arnold).
CURREY, J. D. (1970), *Animal skeletons* (Edward Arnold).
GORDON, J. E. (1968), *The new science of strong materials* (Penguin Books).
GRAY, J. (1968), *Animal locomotion* (Weidenfeld and Nicolson).
HALSTEAD, L. B., and MIDDLETON, J. A. (1972), *Bare bones—an exploration in art and science* (Oliver and Boyd).
HAM, A. W. (1969), *Histology* (Lippincott; Blackwell).
KERMACK, D. M., and KERMACK, K. A., eds. (1971), *Early mammals* (Academic Press).
PENNYCUIK, C. J. (1972), *Animal flight* (Edward Arnold).
SCIENTIFIC AMERICAN (1972), *Vertebrate adaptations* (Freeman).
SMITH, J. M. (1968), *Mathematical ideas in biology* (Cambridge University Press).
WELLS, C. (1964), *Bones, bodies and disease* (Thames and Hudson).
YOUNG, J. Z. (1957), *Life of mammals* (Oxford University Press).

classification of the vertebrates

Phylum **CHORDATA**

Subphylum VERTEBRATA or CRANIATA

Superclass AGNATHA

Class DIPLORHINA
Subclass HETEROSTRACI (PTERASPIDES) *Astraspis, Eriptychius*

Class MONORHINA
Subclass HYPEROARTII
Superorder OSTEOSTRACI (CEPHALASPIDES)
Superorder PETROMYZONIDA Lamprey $\Big\}$ Cyclostomes
Subclass HYPEROTRETI (MYXINI) Hagfish

Superclass PISCES

Class ACANTHODII
Class PLACODERMI Arthrodires, Antiarchs

Class CROSSOPTERYGII
Subclass RHIPIDISTIA
Subclass ACTINISTIA (COELACANTHI)

Class DIPNOI

Class ACTINOPTERYGII
Subclass CHONDROSTEI $\Big\}$ Ganoid fish
Subclass HOLOSTEI
Subclass TELEOSTEI

Class SELACHII Sharks, Rays, Skates

Class HOLOCEPHALI *Chimaera*

Class AMPHIBIA
Subclass LISSAMPHIBIA Frogs, Toads

Superclass REPTILIA

Class EUSAUROPSIDA
Subclass CHELONIA Turtles, Tortoises
Subclass LEPIDOSAURIA Lizards, Snakes
Subclass ARCHOSAURIA Crocodiles, Dinosaurs

Class ICHTHYOPTERYGIA Ichthyosaurs

Class SAUROPTERYGIA Plesiosaurs

Class EOTHEROPSIDA
Subclass PARAMAMMALIA Pelycosaurs, Therapsids

Superclass HOMOTHERMA

Class PTEROSAURIA *Pteranodon, Sordes*

Class AVES
    Subclass SAURIURAE *Archaeopteryx*
    Subclass ORNITHURAE
      Superorder RATITAE
      Superorder CARINATAE Pelican, Albatross, Vulture

Class MAMMALIA
    Subclass PROTOTHERIA
      Order MONOTREMATA Platypus

    Subclass THERIA
      Infraclass METATHERIA
      Superorder MARSUPIALIA
      Infraclass EUTHERIA Placentals
        Order INSECTIVORA
        Order CHIROPTERA Bats
        Order PRIMATES Lemurs, Gorilla, Man
        Order EDENTATA Sloths, Ant-eaters
        Order RODENTIA
        Order LAGOMORPHA Rabbits, Hares
        Order CETACEA Whales
        Order CARNIVORA Bears, Cats, Dogs, Badgers
        Order TUBULIDENTATA Aardvark
        Order PROBOSCIDEA Elephants
        Order PERISSODACTYLA Rhinoceros, Horse
        Order ARTIODACTYLA Giraffe, Pig, Cattle, Sheep

INDEX

calcium phosphate 13, 19, 24, 32, 33, 37, 39, 106, 167
calcium triphosphate 39
calcospherite 75
calculus 167, 169
callus 53, 159
camber 154, 155, 157
canaliculae 46, 59, 64, 79, 80
canals of Williamson 70, 85
cancellous bone 58, 62, 65, 164
canine 136
cantilever 142
capiti mandibularis 123, 124
cap stage 86
carbohydrate 10, 15
Carboniferous 123
caribou 99, 146
carnassial teeth 136
carnivore 90, 117, 118, 127, 136, 144, 146
cartilage 10, 17, 33, 39, 43, 51, 54, 56, 57, 65–67, 102, 159, 163
cartilaginous bone 65
cartilaginous fish 7
cartilaginous joint 119
cat 96, 127
cattle 72, 98
cave-bear 166
C axis 26
C cells 28
C cusp 131, 133
cementing horn 101
cementocyte 71
cementum 60, 71, 72, 118, 136
cephalaspid 56, 79, 107
cerebellum 154
cervix 10
Cervus 99
cetacean 118
chameleon 96
cheetah 144, 146
chicken 10
Chimaera 71
chondroblast 43, 51, 159
chondrocyte 51, 53, 54, 57, 66, 120, 159
chondroitin sulphate 16, 17, 51, 53, 55, 103, 120
clavicle 153, 154
claw 17, 19, 95, 98
cod 10
codon 2
collagen 7–11, 13, 14, 19, 23, 24, 26, 29, 34, 35, 39, 42, 43, 45, 47, 51, 53, 55, 59–62, 64, 70, 75, 111, 112, 114, 115, 118, 120, 154, 164
compression 138

cone layer 103, 104
contour feather 97
Cook-Gordon crack stopper 113, 114
coracoid 153, 156
cornea 10
coronoid bone 123
coronoid process 117, 124
cortex (bone) 65–68, 112, 122, 138, 161
cortex (hair) 94
cosmine 80, 84, 85
cosmoid scale 84
cow 128, 136
Cretaceous 7, 93, 133, 134
cretin 163
crocodile 96, 104
crossopterygian fish 5, 84
cross-ties 139
crosswise struts 154
crystallites 26
ctenoid (scale) 85
cuticle (hair) 94
cycloid (scale) 85
cyclomorial 82, 84
cyclops 162
cysticercus 169
cystine 21
cytosine 1

deer 144
dental caries 167
dental epithelium 87
dental lamina 86
dental papilla 82, 86, 87
dentary 124, 125
dentine 10, 39, 42, 61–63, 73–77, 79–82, 84, 85, 90, 102, 136
deoxyribonuceic acid (DNA) 1, 35
dermis 94
Devonian 7, 8
diaphysis 65–67, 163
diastema 129
digastric 126, 128, 129
Dimetrodon 123
dinosaur 7, 67, 69, 72, 104, 105, 139, 141
Diplodocus 139
disaccharide 16
DNA 1, 35
dog 115, 146
dolphin 137
down (feather) 97
drag 147, 149
dugong 163
duplication (gene) 5, 7, 9
dynamic soaring 150

174

177

178

thixotropy   21, 121
threonine   8–10
thymine   1, 2
thyroid   28
tine   99
Tomes process   87–89, 93
tonic   126
tonofilaments   87
toothed whale   90, 137
torsion   138
tortoise   103
tortoise-shell   17, 19
trabeculae   40, 58–62, 65, 121, 122, 139,
    159, 161, 163, 164
transfer RNA   2, 35
translocation (gene)   5, 7, 9
trauma   161, 167
trez   99
Triassic   93, 125
tricalcium phosphate   34, 35
tribosphenic molar   130, 131, 133, 134
trigon   136
trigonid   133
tRNA   2, 35
tropocollagen   10, 12–15, 26, 42
tubulidentate   80
turbulence   147
turtle   103, 105

ulnar sesamoid   154
ultimobranchial glands   26
uncinate process   153

ungulate   90, 100, 144
uracil   2
urine   24, 27, 29
uterus   10, 103
utriculus   106

vane (feather)   97
vasodentine   80
vasotocin   105
"velvet"   99
vertebrae   141, 142, 153
vitamin C   12
vitamin D   27, 28, 163
vitrodentine   82
Volkmann's canals   68
von Korff's fibres   74
vulture   149, 150

whale-bone   17, 19, 101
whisker   19
wing   147, 149, 156, 157
wing loading   149
wishbone   153
wool   17
woven bone   67

xylose   16

Young's modulus of elasticity   111

zygomatic arch   126

179

THE WYKEHAM SCIENCE SERIES

THE WYKEHAM TECHNOLOGY SERIES

All orders and requests for inspection copies should be sent to the appropriate agents. A list of agents and their territories is given on the verso of the title page of this book.